Matthias Hofherr

Real-time imaging systems for superconducting nanowire single-photon detector arrays

Karlsruher Schriftenreihe zur Supraleitung   Band 016

HERAUSGEBER

Prof. Dr.-Ing. M. Noe
Prof. Dr. rer. nat. M. Siegel

Eine Übersicht über alle bisher in dieser Schriftenreihe
erschienene Bände finden Sie am Ende des Buches.

Real-time imaging systems for superconducting nanowire single-photon detector arrays

by
Matthias Hofherr

Dissertation, Karlsruher Institut für Technologie (KIT)
Fakultät für Elektrotechnik und Informationstechnik, 2014
Hauptreferent: Prof. Dr. rer.nat. habil. Michael Siegel
Korreferent: Prof. Dr.-Ing. habil. Hannes Töpfer

Impressum

Karlsruher Institut für Technologie (KIT)
KIT Scientific Publishing
Straße am Forum 2
D-76131 Karlsruhe

KIT Scientific Publishing is a registered trademark of Karlsruhe
Institute of Technology. Reprint using the book cover is not allowed.

www.ksp.kit.edu

Print on Demand 2014

ISSN 1869-1765
ISBN 978-3-7315-0229-6
DOI 10.5445/KSP/1000041279

Real-time imaging systems for superconducting nanowire single-photon detector arrays

zur Erlangung des akademischen Grades eines

DOKTOR-INGENIEURS

von der Fakultät für
Elektrotechnik und Informationstechnik
des Karlsruher Instituts für Technologie (KIT)

genehmigte

DISSERTATION

von

Dipl.-Ing. Matthias Hofherr

geb. in Ludwigsburg

Tag der mündlichen Prüfung: 11.3.2014

Hauptreferent: Prof. Dr. rer.nat. habil. Michael Siegel

Korreferent: Prof. Dr.-Ing. habil. Hannes Töpfer

Nomenclature

τ_{el}	electrical relaxation time
τ_{ep}	electron-phonon scattering time
τ_{es}	heat escape time to the substrate
d	film thickness
I_B	bias current
I_C	critical current
j_s	current density
L_{kin}	kinetic inductance
n_s	density of superconducting charge carriers (Cooper pairs)
R_0	readout impedance
$R_{det}(t)$	time dependent resistance of the normal domain after a photon absorption event in the meander
SDE	system detection efficiency, considers the effective detection efficiency including all losses in the measurement system
T_C	critical temperature
TTS	transit time spread, is identical to the time jitter t_j
ABS	absorptance, describes the absorption of a film including geometric meander parameters
AFE	analog front end
CDMA	code-division multiple access
CE	collection efficiency, describes the effectiveness of the coupling of a free-space radiation into the input waveguide (e.g. fiber) of the measurement system
CFD	constant fraction discriminator
DCR	dark count rate
Nb	Niobium
NbN	Niobium Nitrid
OCE	optical coupling efficiency, describes the photon losses between detector area and radiation beam
SNSPD	superconducting nanowire single-photon detector
TaN	Tantalum Nitrid
TCSPC	time-correlated single-photon counting
TDMA	time-division multiple access
ZC	zero crossing level

Kurzzusammenfassung

Der Mensch strebt nach immer tieferem Verständnis seiner Umwelt und versucht die daraus gewonnenen Erkenntnisse sich zu Nutzen zu machen. Dabei bedient er sich mehr und mehr auch technischer Unterstützung womit er natürliche Grenzen überwinden kann und ihn somit neue Erkenntnisse gewinnen lässt. So gehören auch sensitive Detektionssysteme zu den Hilfsmitteln, die in den letzten Jahren mehr und mehr erforscht und entwickelt wurden. Die Herausforderung bei der Entwicklung von sensitiven Detektionssystemen liegt in dem Versuch, die geringen detektierten Informationen von dem Eigenrauschen des aktiven Detektionselementes separieren zu können. Das Eigenrauschen eines Systems ist vor allem durch das thermische Rauschen, auch Johnson-Rauschen genannt, bestimmt. Der entscheidende Faktor liegt hier in der Wahl der Betriebstemperatur und dem aktiven Widerstand des Detektionselementes. Gängige Raumtemperatur-gebundene Systeme nutzen oftmals die Eigenschaft der Nicht-Korrelation des Rauschens aus, um mit ausreichender Beobachtungszeit den Rauschanteil in einem Signal herausmitteln zu können. Dies wiederum schränkt die Geschwindigkeit eines solchen Detektionssystems ein und reduziert die gewonnene Information auf einen zeitlichen Mittelwert.
Supraleitende Detektoren können hier ihren Vorteil ausspielen. Zum einen profitieren sie von der großen Widerstandsänderung beim Übergang von der supraleitenden in die normalleitende Phase. Dadurch sind nur kleine Energien notwendig, um eine große Signaländerung zu erhalten. Sie haben aber auch dank der geringen Betriebstemperatur nahe dem absoluten Nullpunkt ein sehr geringes Eigenrauschen im normalleitenden Zustand. Dies betrifft auch die weiteren resistiven Komponenten im kryogenen Umfeld. Diese Punkte allein machen supraleitende Detektoren bereits zu interessanten Kandidaten auf dem Detektormarkt. Hinzu kommt, dass sie auf Grund ihrer intrinsischen Eigenschaften über eine materialspezifische Energielücke verfügen, die mehrere Größenordnungen unterhalb der Energielücke von Halbleitern liegt, weswegen ihr Einsatzfähigkeit über einen breiteren spektrale Bandbreite ermöglicht werden kann.

Diese Arbeit beschäftigt sich mit der Weiterentwicklung von supraleitenden Einzelphotonen Detektoren (Englisch: Superconducting nanowire single-photon detector (SNSPD)). Das Konzept der SNSPDs beruht auf der Eigenschaft, einen supraleitenden Nanodraht allein durch die absorbierte Energie eines Einzelphotons normalleitend zu bekommen und dadurch ein Spannungspuls zu erzeugen. Das typische Anwendungsfeld liegt im Bereich der optischen und Infrarotstrahlung. Es gibt aber auch Experimente im Bereich der Röntgen- und Ionenstrahlung.

In den letzten Jahren wurde weltweit an der Entwicklung dieser Einzelphotonendetektoren geforscht. Im Vordergrund stand die Entwicklung geeigneter Detektionsmodelle, die Optimierung diverser Detektorparameter, wie zum Beispiel Detektionseffizienz, Dunkelzählraten und Zeitjitter und die Entwicklung spezieller Bauformen. Darüber hinaus wurden mehr und mehr die Detektoren für reale Anwendungen eingesetzt. Typische Anwendungsfelder sind zum Beispiel die Quantenkryptographie und Quantenoptik, die Spektroskopie oder die Laufzeitkamera Entwicklung.

Diese Arbeit teilt sich in mehrere Schwerpunkte auf. Zu Beginn ist eine Übersicht über die gängigen Detektormodelle und die im weiteren Verlauf der Arbeit verwendeten Detektorparameter definiert und an Hand von vorausgegangenen Forschungsarbeiten hergeleitet, bzw. erläutert.

Die Charakterisierung von SNSPDs benötigt speziell auf die einzelnen Parameter optimierte Messbedingungen. Dies beinhaltet zum einen generell den Betrieb bei kryogenen Temperaturen. Dabei sollte eine hohe Anzahl von Messzyklen möglich sein, was mit Hilfe eines Dipstick-Kryostates realisiert wurde. Außerdem nötig war eine generelle Optimierung der elektronischen Auslese auf absolute Rausch- und Störarmut, was mit Hilfe von guter Verstärkerselektion, Filterung und der Eigenentwicklung von Biasstromquellen und hybriden HF-Frequenzweichen (sogenannte BiasT) erreicht wurde. Ein BiasT ist als Frequenzweiche notwendig für die Trennung der innerhalb des Detektors auf der gleichen Leitung vorliegenden dc und ac Komponenten. Eine Integration dieses BiasTs ermöglicht eine höhere Packungsdichte, was zu höherer Pixelanzahl führt.
Für spektral breitbandige Anwendungen wurde speziell ein Kalibrierungsverfahren entwickelt, das mit Hilfe von Multimodefaser gekoppelter nicht-kohärenter Strahlung die Photonenrate auf dem Detektor bestimmt, was für die Messung der Detektionseffizienz notwendig ist. Im Laufe der Arbeit wurde außerdem die Jittermessung im Kryostat optimiert, um hochaufgelöste zeitkorrelierte Einzelphotonenmessungen (TCSPC) durchzuführen. In diesem Zusammenhang wurde auch ein analoges Frontend entwickelt, das mit einer Bandbreite von 10 GHz eine Pulsanpassung für den Eingang typischer TCSPC Elektroniken erlaubt und die Charakterisierung und den Betrieb von SNSPDs erleichtert und verbessert.

Ein großer Forschungsschwerpunkt dieser Arbeit liegt auf der Untersuchung des intrinsischen Verhaltens von SNSPDs bezüglich Detektionseffizienz und Dunkelzählrate. Die Variation von geometrischen Parametern ermöglicht einen direkten Eingriff in die Festkörpereigenschaften der Supraleiter und ist gleichzeitig verhältnismäßig einfach zu realisieren. Daher wird in dieser Arbeit die Abhängigkeit der Detektionseffizienz von der Filmdicke analysiert und in Relation zur Filmabsorption gesetzt, um intrinsische Grenzen der Detektoren zu bestimmen und zu verbessern. Zur Vervollständigung des Verständnisses wurden weitere Analysen wie die Biasstromabhängigkeit und eine Amplitudenanalyse ergänzt. Abgesehen von der Evaluation von geeigneten Fabrikationsparametern hinsichtlich guter Detektorperformanz zeigte sich vor allem in der Analyse der oben genannten Parameter bezüglich der eingestrahlten Wellenlänge, dass für den Infrarotbereich eine anderes Detektionsmodell als das gängige Hotspot Model notwendig ist. Basierend auf den Vorarbeiten unserer Mitautoren aus der Kollaboration mit dem DLR Berlin konnte ein auf Vortexassistenz basierendes Modell anhand der Detektionseffizienzmessungen verifiziert werden. Dieses Vortexmodel wird auch als Ursache für die Dunkelzähler angenommen. In einer dreiteiligen Analyse wurde daher die Abhängigkeit der Dunkelzählrate von der thermischen Ankopplung des Detektors an das Kühlreservoir, der Abhängigkeit von der Betriebstemperatur während der Messung und der Abhängigkeit von der stöchiometrischen Filmzusammensetzung untersucht und daraus resultierende Tendenzen für niedrigere Dunkelzählraten bestimmt. Darüber hinaus wurden die Messungen ebenfalls mit dem Vortexmodel verglichen. Auch hier zeigte sich, dass das Vortexmodel als plausible Beschreibung der Dunkelzählerentstehung in Frage kommt.

Ein weiterer Schwerpunkt in dieser Dissertation ist die Entwicklung von Multipixelsystemen basierend auf SNSPDs. Multipixelanwendungen kann man in mehrere Anwendungsgebiete unterteilen. Es geht entweder um die frameweise Bestimmung mittlerer Zählraten, sogenannte Imaginganwendungen, was für Videosysteme von Interesse ist, oder es geht um die Echtzeitauswertung eines Multipixelchips, bei der das Auftreten jedes Photons analysiert werden kann. Eine dritte Anwendung wäre theoretisch noch die Auswertung von Multiphotonereignissen, was aber nicht im Fokus dieser Arbeit steht.

Für Imaginganwendungen wurde in dieser Arbeit die Verwendung von supraleitender Rapid-Single-Flux-Qantum (RSFQ) Elektronik als Ausleseelektronik untersucht. Anhand einer 4-Pixel Messungen wurde zum einen zum ersten Mal auf der Welt die Machbarkeit einer solchen Auslese gezeigt, es wird aber auch die Skalierbarkeit dieser Elektronik diskutiert. Da die RSFQ Funktionalität im Kern auf eine asynchronen Kanalverknüpfung aller Eingangskanäle beruht, war es nötig, das Verfahren noch mit einem intelligenten Biasing-Konzept basierend auf Time-Division und Code-Division Multiplexing-Verfahren zu erweitern, um die Rekonstruktion der jeweiligen Photonenraten für jedes einzelne Pixel zu ermöglichen.

Für die Echtzeitauslese eines Multipixelchips wurde ein Time-Tagged Multiplexing Verfahren entwickelt. Diese Methode basiert auf der Verzögerung der Detektorpulse, die für jedes Detektorelement individuell festgelegt werden kann. Dadurch kann durch einen Laufzeitvergleich das ursprünglich auslösende Detektorelement der gemessenen Pulsantwort zugeordnet werden. Der Vorteil dieses Verfahrens ist, dass zum einen die Anzahl der Versorgungs- und Ausleseleitungen auf ein Minimum reduziert werden kann, was vor allem der seriellen Anordnung der Detektorelemente zu verdanken ist. Außerdem ist der Fabrikationsaufwand nur geringfügig komplizierter, da anstelle einer separaten kryogenen Elektronik nur ein erweitertes Maskendesign für die Detektorherstellung benötigt wird. Die Auslese erfolgt nach der zeitkorrelierten Einzelphotonen-Methodik (TCSPC), welche einen Referenzpuls zur Zeitbestimmung nötig macht. Diese würde an sich das Anwendungsgebiet dieses Multiplexing-Verfahrens auf gepulste Anwendungen reduzieren. Allerdings konnte in dieser Arbeit gezeigt werden, dass ein Time-tagged-Detektor zweikanalig ausgelesen werden kann, wodurch die dem Detektorelement zugehörige Zeitsignatur intrinsisch aus den unterschiedlichen Laufzeiten der beiden Kanäle bestimmt werden kann, ohne die Notwendigkeit einer gepulsten Anregung.

Contents

1 Introduction

The truth lies behind the limits! In ancient times, the old Egyptians looked at the sky with their eyes. What they saw, were stars as little dots of light. They only realized a certain number of stars, which cover the hemisphere because the eyes were limited in sensitivity. They defined star constellations like the Orion, which they then honored in their monuments [1]. Later, mankind developed techniques that increases the collection of light: The use of lenses in combination with telescopes. However, still the eye, as final "detection system", had a limited sensitivity. Hence, several thousand years after the time of the ancient Egyptians the development of electronic systems opened new ways in detection. These detectors facilitate the great astronomical pictures, to which we are used nowadays, which enable quite more information and understanding about the truth of the universe.

However, in astronomy, but also in other fields of science and engineering like spectroscopy [2], quantum communication [3] etc., a detector sensitivity down to the single-photon regime would increase tremendously the gain in knowledge. Some detector systems, like a pyrometer, use the integration time as sensitivity enforcing parameter by averaging the background noise and therefore, increase the signal-to-noise ratio. However, the sensitivity is then close correlated to the measurement time. An evaluation of single-photon information is not possible. In the semiconductor technology single-photon avalanche diodes (SPADs) and photomultiplier are meanwhile the strongest candidates for single-photon resolution and already quite common in science and industries [4]. However, this type of detectors have certain limitations, e. g. the dead time of avalanche diodes can be quite large in the range of several tens of nano seconds, which decreases the repetition rate of a detection system. The main limitation is the limit in the spectral sensitivity. In a semiconductor, the band gap Δ_{sem} defines the minimum photon energy that is required to detect the photon. Silicon has a band gap of $\Delta_{sem} = 1,1$ eV and its transparency characteristic in the infrared regime strongly limits the spectral bandwidth. Therefore, an additional semiconductor is required and often Germanium ($\Delta_{sem} = 0,67$ eV) is used to cover the infrared range (see section 4.1.3). One can conclude from these facts that it is not possible to cover a wider spectral range by only one semiconducting detector. In principle, multiple-detector systems can solve this problem, but a multichip system leads to a quite complex construction and suffers from the different characteristics of each detector type.

So far, we considered only the single-photon capability of a single-pixel system. For imaging applications and in spectroscopy [2] it would be more interesting to get multipixel systems. Multipixel systems of avalanche diodes already exist [5]. However, the number of pixels is still quite low.

Since Heike Kamerlingh Onnes has discovered the superconductivity in 1911 [6], superconductivity became more and more interesting for applications. Especially the theoretical prediction of the Josephson effect in 1962 by Brian D. Josephson [7], which was later proved in experiment, and the high-temperature superconductivity [8] led to exciting new effects. The use of superconductivity as detection mechanism is obvious, since the strong change from zero-resistance to the normal-conducting case at the critical temperature T_C and the critical current I_C, respectively, is predesignated as detection effect. The superconducting gap Δ_{sup} is in the range of several meV and therefore, much smaller than the semiconducting gap Δ_{sem} [9]. In the Δ_{sup} range, only small activation energies are required to get measurable changes in the order-parameter of the superconductor, which even can be modified by parameters like temperature and the current in the superconductor. The way, the superconductivity can be used as detection mechanism, varies, depending on the type of application [10]. However, they all base on the same superconducting physics [11].

A special type of superconducting detector is the superconducting nanowire single-photon detector (SNSPD). It consists of a very narrow superconducting line and an ultra-thin film, which is operated at cryogenic temperatures. In connection with an applied bias current I_B, which is set close to the critical current I_C , this type of detector can respond to a single-photon event by a short resistive domain in the wire, which leads to a voltage transient.

In the mean time, since the first considerations [11], this type of detector causes a strong research ambition of many groups in the world [12]. SNSPDs gain from an extremely wide spectral sensitivity from the X-ray [13] to the middle infrared regime [14]. It is even used for neutron measurements and other elementary particles [15]. A very low dark count rate and non-comparable low time jitter values down to 18 ps [16] make them interesting for time-correlated single-photon counting (TCSPC) measurements [17]. This strong research has led to competitive detector parameters, making them suitable for many applications in optics, communication and imaging [12].
Up to now, much research has been done to optimize the detection parameters like detection efficiency, dark count rate and dead and jitter times [12]. Nevertheless, there are still open questions that reach deep in the intrinsic processes of the SNSPDs. Even the detection mechanism itself is not fully understood. Answers to these questions would help to find strategies of improvement for individual application tasks.
A multipixel readout is a strong requirement for applicable use. The pure copy of a single-pixel readout concept will fail for high pixel numbers. Here, a conflation of physical under-

standing of intrinsic mechanism and typical engineering tasks is necessary to develop new ideas for compact readout schemes.

In chapter 2, the basic models and parameters of SNSPDs are explained and set in a context with current research. In chapter 3, the self-developed SNSPD readout system is presented and explained in detail. The core of this thesis are two research topics: In chapter 4, the focus is set to an improved understanding of intrinsic detection processes in the SNSPD to improve the performance of an SNSPD. An enlarged understanding of the detection model is based on a systematical analysis of the detection efficiencies dependent on superconducting film thickness. The dark count rate of SNSPDs is optimized by using three modifications: The variation of the thermal coupling, the operation temperature and a film deposition parameter. The second main topic, described in chapter 5, is the introduction and the verification of two multiplexing concepts for multipixel SNSPD readout. One readout concept works with a Rapid Single-Flux Quantum electronics (RSFQ) in combination with multiplexing methods from the field of communication techniques. The world wide first system demonstration was realized. The other, likewise novel concept uses a time delay as multiplexing method, which is evaluated by a time-correlated single-photon counting (TC-SPC) measurement. This concept requires only a modified patterning of the detector chip and reads out an array of detectors by one readout line.

2 Superconducting nanowire single-photon detector (SNSPD)

In this chapter the typical structure and operation principle of an SNSPD is introduced. To understand later analysis tasks the intrinsic physical processes of an SNSPD are explained. Moreover, several detector parameters are revealed which characterize the performance of an SNSPD. An overview about the current state of worldwide research is given for each of these parameters.

2.1 The principle of an SNSPD

2.1.1 The hot-spot model and electro-thermal model

The photon energy of optical and infrared photons is in the range of several eV. The absorbed energy in the SNSPD has to be larger than the superconducting energy gap Δ_{sup}, which in usually in the range of few mV [9]. However, the energy gap of a semiconductor Δ_{sem} is more than one order of magnitude larger than Δ_{sup} of a superconductor. Therefore, superconducting detectors have a broader spectral range than semiconducting detectors. Δ_{sup} depends on the superconducting material [18] but can be influenced during operation e.g. by the bias current [19] and the temperature [EIS$^+$13].

The elementary structure of an SNSPD is a nanowire with a nominal width of 100 nm made from a thin layer of a superconductor (see Fig. 2.1). Typical superconductor materials for these detectors are NbN [20] , Nb [21] , TaN [EIS$^+$13] , NbTiN [22] and amorphous tungsten silicide [23]. Various other materials might be considered for SNSPDs, as well. To increase the effective area of the detector, the nanowire is typically patterned in a meander shape. Further modifications of this basic design, like parallel stripes [24] or spiral detectors [HRD$^+$13], are meanwhile presented to improve certain performance parameters, as well. In this thesis, NbN is used as superconductor for SNSPD devices due to its high critical temperature T_C.

The detection mechanism of an SNSPD is called hot-spot mechanism (see scheme in Fig. 2.2). An SNSPD is operated by applying a bias current I_B close to the critical current I_C. The critical current I_C itself depends on various parameters like e.g. temperature T [18], [HRI$^+$10b], and linewidth w. The bias current I_B acts as a reduction of the superconducting order parameter. The energy portion $\delta\Delta$ which is still required to switch the wire normal-

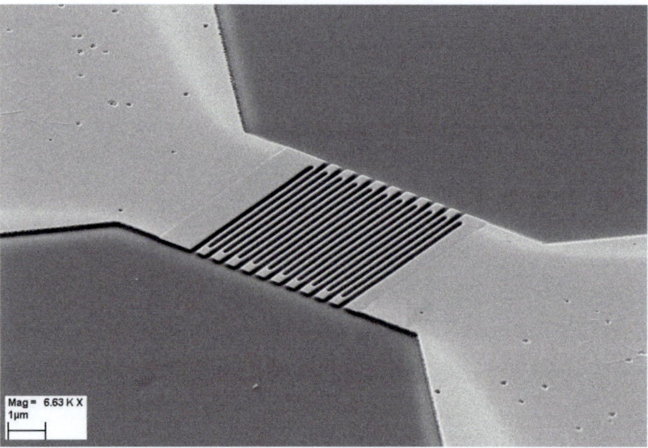

Figure 2.1: Typical geometry of a superconducting meander.

conducting is in the range of the absorbed photon energy. The current density j_s of the bias current is defined by:

$$j_s = 2e \cdot n_s \cdot v_s \qquad (2.1)$$

A change of the superconducting order parameter due to an absorbed photon changes the number of the superconducting charge carriers n_s by breaking Cooper pairs.

$$n_s' = n_s - \delta n_s \qquad (2.2)$$

The Cooper pair breaking can also be considered as an increase of the local temperature, equal to a local non-equilibrium, and therefore, it is called a hot spot. However, the bias current, which is applied by the external bias source, stays constant ($I = 2e \cdot n_s \cdot v_s \cdot w \cdot d =$ const). The charge carrier velocity v_s follows reciprocally any change in the number of

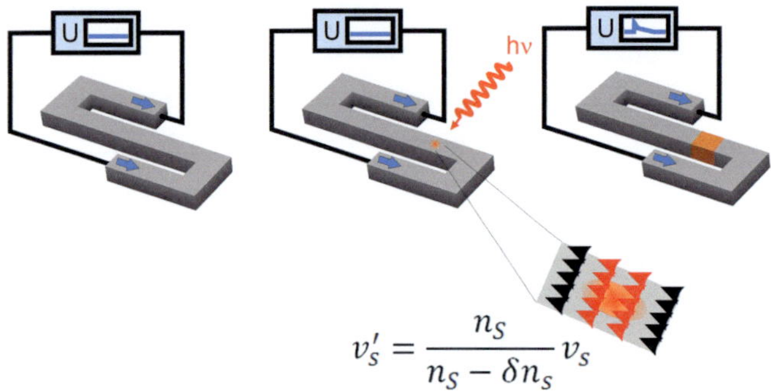

$$v_s' = \frac{n_s}{n_s - \delta n_s} v_s$$

Figure 2.2: Scheme of the hot-spot detection mechanism of an SNSPD.

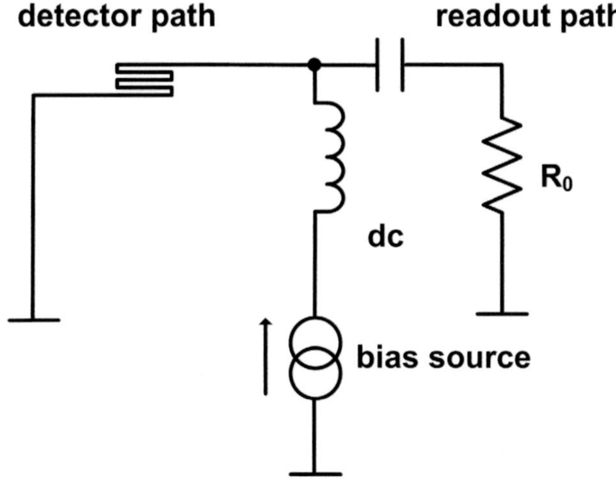

Figure 2.3: Simplified circuit of an SNSPD readout

charge carriers n_s [25]. This is a simplification because this direct proportionality between n_s and v_s is only valid for changes slower than the relaxation time τ_{ep} [26].

$$v_s' = \frac{n_S}{n_s - \delta n_s} \cdot v_s \tag{2.3}$$

If the velocity reaches a critical value $v_s = v_{s,c}$, a small section of the current carrying wire becomes normal conducting in the range of twice the coherence length ξ. This normal belt growths and shrinks depending on the current $I_{det}(t)$ and the temperature T in the meander, which is described by the electro-thermal model. The voltage transient is there defined by the differential equation [20]:

$$-L \cdot \frac{dI_{det}}{dt} + R_{det}(t) \cdot I_{det}(t) = R_0 \cdot I_{readout}(t) \tag{2.4}$$

$$I_B = I_{det} + I_{readout}$$

L is the inductance of the detector meander. The bias current I_B in the detector meander is directly affected in the moment of the occurrence of a normalconducting belt, as well. Due to the fact that the impedance of the readout line acts as a parallel resistance to the detector, the current I_B is split in the remaining current I_{det} in the detector and the current $I_{readout}$, which is redistributed from the detector path to the readout path [20] (see Fig. 2.3).

Although the voltage and current distribution in the detector system can differ on each electrical component, the measured response amplitude refers to the voltage on the readout impedance R_0 and is defined by $U(t) = R_0 \cdot I_{readout}(t)$. Fig. 2.4 shows a typical response measured at room temperature.

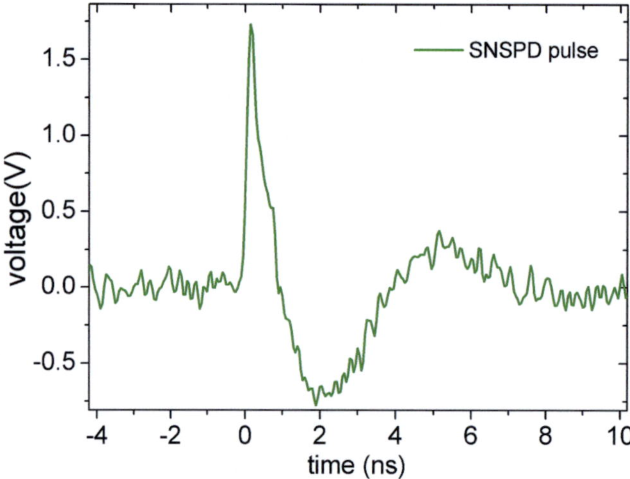

Figure 2.4: Typical amplified detector response of a 4 nm thick NbN SNSPD at 6 K.

However, in all cases the normal domain $R_{det}(t)$ shrinks due to the heat transport to the substrate until it vanishes completely. The re-flow of the bias current in the detector path is hampered by a LR network, which defines a relaxation time constant. The time constant of this process is given by $\tau_{el} = L_{total}/R_{total}$. L_{tot} is the total inductance of the detector and readout path, and is mainly defined by the inductance of the detector, and R_{tot} is the resistance of all resistive parts in the total signal path. Since one assumes that $\tau_{el} > \tau_{es}$ (τ_{es} roughly defines the time during which the normal domain exists in the meander (see section 2.1.2), R_{det} is already zero again during most of the decay time of a detector pulse response. Therefore, most calculations to define L_{tot} from the measured τ_{el}, are done with $R_{tot} = 50\ \Omega$, which is given by the impedance $R_0 = 50\ \Omega$ of the readout line. [27].

2.1.2 Two-temperature model of superconductors

If a photon is absorbed in the SNSPD the photon energy causes a certain non-equilibrium. To reach a stationary state again, the photon energy has to dissipate out of the detector system. This energy relaxation is defined by dynamic processes of two separated reservoirs in the superconductor: The electrons or – as defined for superconductors – the quasi-particles, and the phonons. Both reservoirs can have different temperatures T due to a certain decoupling. The dynamic process of the non-equilibrium state can be described by two differential equations (see Eq. 2.5), which are called the Two-Temperature Model [28], [11].

$$c_e \frac{\mathrm{d}T_e}{\mathrm{d}t} = -\frac{c_e}{\tau_{ep}} (T_e - T_p) + \alpha P(t)$$

$$c_p \frac{\mathrm{d}T_p}{\mathrm{d}t} = -\frac{c_e}{\tau_{ep}} (T_e - T_p) - \frac{c_p}{\tau_{es} (T_p - T_0)}$$

$$(2.5)$$

The parameters c_e and c_p describe the electron and phonon specific heat capacity. The relaxation time constant τ_{ep} is the electron-phonon interaction time and τ_{es} is the phonon escape time to the substrate. T_e and T_p are the two reservoir temperatures, which, by the difference $T_e - T_p$, define the relation between the two equations. $P(t)$ is the inserted photon power. The factor α is the absorption coefficient of the film. The absorbed photon energy heats the electron reservoir and increases T_e. This process described by the electron-electron scattering time τ_{ee} is quite fast ($\tau_{ee} < 1ps$) and neglected in this context. The electron-electron scattering process is followed by a scattering of the energy to the phonon sub-system, which represents a heating of the atomic lattice. The scattering from the electrons to the phonons given by τ_{ep} is higher than the back-scattering time ($\tau_{ep} >> \tau_{pe}$) due to $c_e/c_p < 1$. Therefore, an equilibrium between the temperature of the electron and the phonon reservoir is not possible in this moment. In literature exist several approaches to define τ_{ep}. For NbN a value of 7 ps is given in [29]. In comparison, in [30], a value of 15 ps is published and in [11] a value of 17 ps. However, [30] describes already the total time of the hot-spot evolution process. Although this process is very fast, it is already in a range which is close to important detector parameters like the detector jitter t_j (see section 2.2.4).

In the two-temperature model, the heat dissipates along the atomic lattice to the substrate. This relaxation process is described by the relaxation constant τ_{es} . Due to the spatial heat transport, the relaxation τ_{es} is depending on the thickness d of the film: $\tau_{es} = \frac{d}{\alpha v_s}$. v_s is the speed of phonons in the lattice. The transition between superconducting film lattice and substrate lattice depends on the lattice constants of the materials used. The quantity of reflection-free lattice vibrations on this interface is described by α [31], [RPH$^+$10]. A value of τ_{es} for NbN is 115 ps [RPH$^+$10].

2.1.3 Voltage response of a superconductor to absorbed energy

Up to now, we discussed the intrinsic heat transport in the superconductor by its relaxation constants, since they define the maximum speed of energy transport. The measurable output signal of a detector is a voltage transient, whose evolution is not described so far. Therefore, it is required to show the conversion between the incoming energy, the intrinsic heat processes and the output voltage signal. From an engineer's point of view, this conversion can be specified by a transfer function, with a certain frequency, or time dependence, respectively. In case of a superconducting detector one can define the transfer function as the bolometric sensitivity S, which gives the relation between the input radiation $p(t)_{ap} \circ\!\!-\!\!\bullet\, P_{ap}$ and the output voltage signal $u(t)_{det} \circ\!\!-\!\!\bullet\, U_{det}$.

$$U_{det} = S \cdot P_{ap} \tag{2.6}$$

The sensitivity S in the frequency domain of a superconducting film can be derived from the two-temperature model, numerically solved by [32], and analytically described by [33]:

$$S = \frac{\alpha I \, dR/dT \, \tau_{es}}{(c_e + c_p) \cdot V} \cdot \sqrt{\frac{1 + \left[\omega \tau_{ep}\left(1 + \frac{c_p}{c_e}\right)\right]^2}{\left[1 + (\omega \tau_{es})^2\right] \cdot \left[1 + (\omega \tau_{ep})^2\right]}} \qquad (2.7)$$

Eq. 2.7 describes the norm of the transfer function, without the phase information. The first term describes the basic bolometer equation and the term in square root the dynamic evolution over time [34], which contains the relaxation constants from the two-temperature model.

Eq. 2.7 is valid if the system is operated near the superconducting/normalconducting transition and a certain current I is applied to the superconductor. Eq. 2.7 is only valid for small deviations from the thermal equilibrium. The differential resistance dR/dT is a linearization of the transition resistance in the thermal operation point. The time transient response $u(t)_{det}$ of such a system is the inverse Fourier transform (IFFT) of $S \cdot p_0$ (if one assumes that $p(t)_{ap} = p_0 \cdot \delta(t - t_0)$).

The mathematical interpretation of this equation can be given as follows: The frequency-dependent expression of Eq. 2.7 has two poles and one zero point. Each pole defines the 3 dB-cutoff frequency of a low pass. The zero point in the numerator defines a shift of the decay in the frequency domain because the zero point can diminish the low-pass behavior of a pole which leads to an increase of the frequency bandwidth of the detector and consequently the detector speed. The zero point is dependent on τ_{ep}, but also on the ratio c_p/c_e. The ratio is material dependent. In NbN this relation is $c_p/c_e = 6$, [29]. The zero-point frequency is close to the 3 dB cutoff frequency in the denominator, which nearly leads to a cancellation of these terms. Therefore, the frequency bandwidth of a NbN detector is always good but not as good as other superconducting detectors like made from YBCO, which has $c_p/c_e = 38$ [TRS$^+$13].

An SNSPD is operated quite underneath the resistive transition of the superconductor. Therefore, S can not be used directly to describe the SNSPD voltage response. The SNSPD response results from both the two-temperature model as described in Eq.2.5 and a electro-thermal model [27], [35], [20], which is defined by the interaction of the detector and the surrounding readout circuit. However, the relaxation constants in the frequency term define the minimum pulse scale that the SNSPD response can reach.

A detector can not be faster than defined by its material specific relaxation constants (τ_{es} and τ_{ep}) and specific heats c_e and c_p. Semenov defines in [10] an absolute limit by a common relaxation constant τ_ε to:

$$\tau_\varepsilon = \tau_{ep} + \left(1 + \frac{c_e}{c_p}\right) \cdot \tau_{es} \qquad (2.8)$$

τ_{ε} can be taken as estimation of the time span during which a normal resistance R_{det} exists in the SNSPD meander after a photon absorption.

2.2 Typical detector parameters and ongoing state of research

In the following section several SNSPD performance parameters like detection efficiency, dark count rate and the transit time spread are explained. Typical values of these parameters are discussed in context with various experimental results from worldwide research groups. Additionally, the kinetic inductance, which plays a major role in the evolution of the SNSPD response, is explained. Moreover, a short overview about the current state of alternative multi-pixel readout and spectroscopy application with SNSPD is given.

2.2.1 Detection efficiency and absorptance

A basic definition of the detection efficiency of a detector considers the ratio between the number of counted photons divided by the number of incoming photons.

$$\text{Detection efficiency} = \frac{\text{counted photons}}{\text{incoming photons}} \tag{2.9}$$

The value is always ≤ 1.

This description is too simple to discuss different parameters, which influence the detection efficiency. For this reason a more detailed definition is required [HRI$^+$10b]. A global description of the detection efficiency is the system detection efficiency (SDE), which can be classified in several parts.

$$SDE = OCE \cdot ABS \cdot IDE \tag{2.10}$$

The SDE is defined by the optical coupling efficiency (OCE), the absorptance (ABS), and the intrinsic detection efficiency (IDE). The OCE describes the losses due to the bad ratio between the beam spot at the fiber end or the collimation diameter of a free space beam and the area of the active detector element and all attenuation in the optical path. The IDE is the intrinsic detection efficiency, which is determined by the intrinsic photon-detection mechanism of the superconductor.

The absorptance ABS is derived from the absorption of a material. The absorption of the optical power P in a material is defined by an exponential damping [36]:

$$P(x) = P_0 \cdot e^{-\kappa x} \tag{2.11}$$

The parameter κ describes the material depending absorption coefficient. The ABS is a factor emanating from κ, which takes several further geometric factors into account, as well. The ABS of the meander includes the thickness, width and filling factor (FF) of the meander

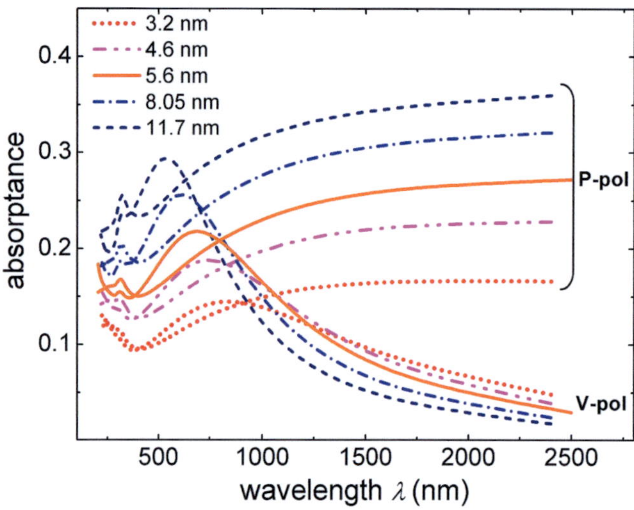

Figure 2.5: Simulated absorptance by S-matrix technique of a NbN meander for different thicknesses, wavelengths and polarizations based on typical NbN film and geometry parameters. [HRI$^+$10b]

and depends strongly on the polarization and wavelength of the incident light [37],[38],[39]. The *ABS* of the SNSPD meander can be experimentally evaluated [37], but it is a challenge to match the focal spot to the squared meander, which increases the measurement uncertainty and make correct analysis of the data difficult. However, a simulated ABS based on real film parameters can approximate the real absorptance of a detector. The *ABS* per unit area of an infinite grating at vertical radiation is computed with the S-matrix technique by Gippius [40]. The permittivity parameter is extracted from a measurement of a non-structured NbN film which is determined at room temperature by spectroscopic ellipsometry [39]. The absorptance is approximated by a simulation based on a wire-width of 90 nm and a filling factor of 50 %. The influence of the contact pads and the readout lines on the *ABS* of the meander are not considered, since these geometries have no influence to the active area of the detector (the active area is bounded by meander lines.). These areas will not respond to a photon absorption. In Figure 2.5 the dependence of the *ABS* on wavelength simulated for different wire thicknesses and both polarizations – parallel and vertical to the meander lines – is demonstrated. The absorptance lies always underneath 35 %. Thicker films have, as expected, a higher absorptance than thinner films. Typical NbN SNSPDs with thicknesses of 4 nm reach only 15 to 20%. Generally, one can say that parallel polarized light leads to higher *ABS* in the wavelength range between 500 nm and 2500 nm [HRI$^+$10b].

Based on the same simulation data, the influence of the filling-factor on the *ABS* is simulated. In Fig. 2.6 the results are given. A higher *FF*, which means that the NbN meander width dominates the gap width between the meander lines, leads to an improved absorption.

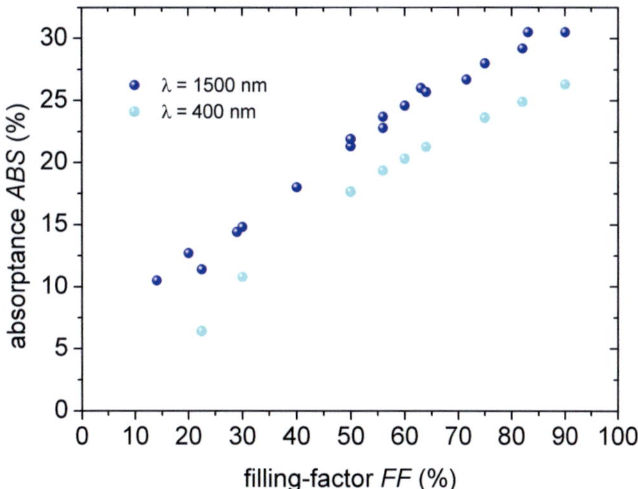

Figure 2.6: Simulated percental absorptance depending on the filling-factor of the meander structure. The simulation is realized for a 90 nm wide meander from a film thickness of 5.6 nm.

Moreover, the absorptance follows linearly the change of the FF in the optical and infrared range, however, slightly shifted. The infrared radiation leads to around 5 % more ABS.

Since the ABS is a direct factor of the SDE, this simulated values shows that the ABS is a high limiting parameter in regard to getting a high SDE.

In literature there exist several approaches to increase ABS. One can distinguish two methods. One focus is on the increase of ABS by varying film parameters like stoichiometry [HDH+12] or the change of the superconducting base material, e.g. the use of Tantalum instead of Niobium [IHR+11] for which values of around > 20 % can be achieved. The other concept bases on the idea to increase the absorption length for a certain wavelength range by either reflecting the incoming radiation on a cavity structure [41],[42] or using optical waveguides to guide the radiation along the absorbent material, which leads to values close to 100 % [16].

In most of the experiments in this thesis, the knowledge of the OCE is not necessary for discussion. Therefore, only the product $ABS \cdot IDE$ will be referred to as the detector detection efficiency DE.

When considering a fiber-coupled system one has to keep a further factor in mind – the so-called collection efficiency (CE) . The CE describes the amount of light that can be collected at the input of the fiber at room temperature. If the light source is already fiber-based as in laser systems, this value can be close to 1. If the system uses free-space optics, the coupling into the fiber can be drastically decreased depending on the optical beam conditions of the source. The definition of an application detection efficiency (ADE) would then be:

$$ADE = CE \cdot SDE \qquad (2.12)$$

With this parameter the efficiency of the complete optical detector system can be described. That means, if an application requires a certain detection efficiency for analysis, the *ADE* value is required to define a detector system as suitable or not.

2.2.2 Dark counts

The performance of an SNSPD is influenced by noise impacts. One has to differ between dark counts that are triggered by the analog noise floor, which superposes the detector signal, and dark counts, which look like real photon counts. In the first case, the noise impact can be weighted by the relation between signal amplitude and background noise, which defines the signal-to noise ration (SNR) of the detector. Depending on the readout electronics, a too low SNR increases the error detection rate, which means that noise peaks are interpreted as photon pulses. In Fig. 2.7 one can see that the detector counts have to be larger in amplitude U_{sig} than the noise peaks (striped area) to get counted correctly. The noise background is produced by the electronic components in the rf and dc path of the measurement setup (see chapter 3). The signal amplitude of an SNSPD response is mainly defined by the bias current I_B. A higher I_B leads to higher pulse amplitudes U_{sig}. Measurements have shown that an SNSPD delivers a sufficient amplitude with $I_B > 10\mu A$ for most of the room-temperature recording electronics used at Micro- and Nanoelectronic Systems (IMS).

The other type of dark counts looks like normal, photon-induced voltage pulses. They can not be filtered out of the response signal, except in case of pulsed excitation[43], [44],

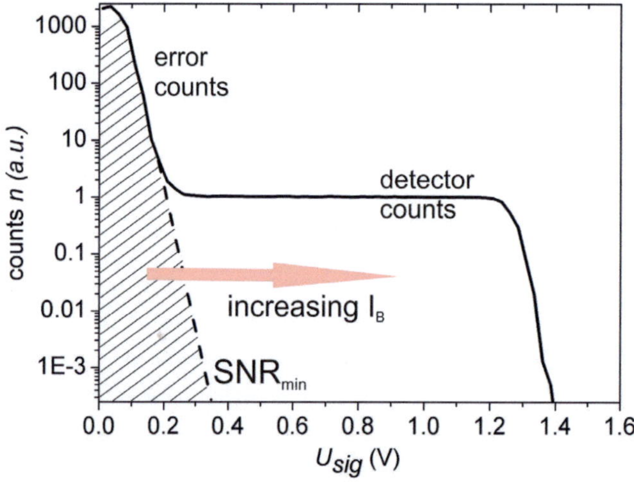

Figure 2.7: Behavior of count rate depending on signal amplitude. The detector amplitude has to be larger than a certain noise level to dominate a possible error detection rate.

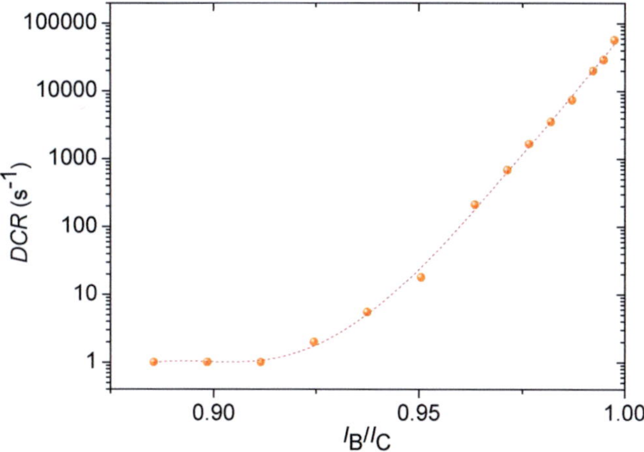

Figure 2.8: Dependence of the dark count rate on the bias current I_B.

since this intrinsic dark counts in this operation mode are not synchronous to the excitation pulse. In all other excitation modes an average dark count rate (DCR) could be defined by closing the radiation input and measuring with the same operation conditions as in the actual optical measurement. This average DCR can be used to calibrate the optical measurements. During the last years, much time was spent to analyze the origin and evolution of intrinsic dark counts. Generally, these dark counts can originate from real photons, caused by black body radiation in the measurement system, which exists due to different temperatures along the optical input path.

However, most of them originate from fluctuations in the film. Their appearance is enforced the closer $I_B/I_C = 1$. In Fig. 2.8 one can see the typcial dependence of the dark count rate on the bias current I_B. The DCR is exponentially growing with the bias current. The starting point and the strength of growth is depending on several parameters like temperature, geometry, film parameters. There are several models to describe these fluctuations: phase-slips [45], thermal excitation of magnetic vortices, [46], [47], [48] and fluctuations in the density of the superconducting condensate [46]. Moreover, there exist investigations that these fluctuations originate enforced at local weakpoints on the meander geometry, e.g. in the meander turns [49].

A typical detector DCR can hardly be defined because the number of dark counts strongly depends on the bias current I_B. Values of less than 100 s^{-1} are often reported, however, one can also find operation points of the SNSPD with a $DCR < 1\ s^{-1}$.

2.2.3 Kinetic inductance of superconducting nanowires

From basic electrodynamic physics it is known that each thin current carrying wire loop possesses an inductance L. This inductance L is defined from physical consideration as:

$$L = n \cdot \frac{\Phi}{I} \tag{2.13}$$

L defines the relation between an applied current I in the wire loop, which generates a magnetic field around the wire and the resulting magnetic flux $\Phi = \oint B dA$ in the cross area A of the loop, n is the number of windings of the wire. Normally, in normalconductors only a geometric/magnetic inductance $L = L_{geo}$ is considered. However, each narrow line has a second type of inductance, as well, the so-called kinetic inductance L_{kin}. L_{kin} is caused by the inertness of the electrons to changes in the velocity. L_{kin} exists in normal conductors, as well. However, the velocity of the electrons in normal conductors is more dominated by the scattering processes between electrons, phonons and atomic defects (represented by the electron mean free path), than by the inertness of the electrons. Consequently, L_{kin} can be neglected in normalconductors. In superconductors, the Cooper pairs have the characteristic of Bosons. As a consequence, there are no scattering processes with the phonons and the atomic defects [18], which enforces the kinetic influence. A typical description of the current flow in a superconductor is given by the two-fluid model, which supposes that in a superconductor a certain density of carriers consists of quasi particles and a certain density consists of Cooper pairs. The ration between these densities is given by the superconducting order parameter. Due to the fact that the resistance in a superconductor is zero, the complete current transport is realized by the Cooper pairs [50].

There exist several derivations of L_{kin} in literature. For the basic understanding of L_{kin} the following definition is suitable derived from the London equations [26].

L_{kin} can be defined by equalizing the kinetic energy E_{kin} of the superconducting charge carriers n_s in a certain volume with the energy E_L stored in an inductance L with applyied current I.

$$E_{kin} = E_L$$
$$\frac{1}{2} \cdot 2 \cdot n_s \cdot m_s \cdot v_s^2 \cdot l/wd = \frac{1}{2} \cdot L_{kin} \cdot I^2 \tag{2.14}$$

Here, m_s is the mass of cooper pairs, w the line width l the line length and d the line thickness. The kinetic inductivity L_{kin} can be extracted by Eq. 2.14 and $I = 2 \cdot e \cdot v_s \cdot n_s \cdot wd$:

$$L_{kin} = m_s/n_s \cdot e^2 \cdot l/wd = \mu_0 \lambda_0^2 \cdot l/wd \tag{2.15}$$

μ_0 is the vacuum permeability and λ_0 the London penetration depth. In this equation, a fixed number of charge carrier n_s is assumed. Taking into account that actually n_s is current and temperature dependent, this definition is only valid for bias currents $I_B = 0$ and temperature T far below T_C, where the Cooper pair breaking is still negligible [26].

The change of L_{kin} with bias current I_B can be approximated [47].

$$\delta L_{kin}(I_B) = \left[\frac{\lambda_0(T,I_B)}{\lambda_0(T,0)}\right]^2 = \left\{1 - 0.31\left[\frac{I_B}{I_C(T)}\right]^{5/2}\right\}^{-1/3} \tag{2.16}$$

The total inductance of a thin superconductor is:

$$L_{total} = L_{geo} + L_{kin}. \tag{2.17}$$

The geometric parameters of SNSPDs are in a regime where L_{kin} dominates L_{total} [27], [20] with the factor $\frac{\Lambda}{w}$ ($\Lambda = \frac{2\lambda_0^2}{d}$, d is the superconducting film thickness) [26]. A certain value of L_{total}, respectively L_{kin} is required in an SNSPD to guarantee latching-free operation of the detector, [51]. Moreover, L_{total} defines the decay length of the SNSPD pulse response [20], [27].

2.2.4 Transit-time spread of SNSPDs

The transit-time spread (TTS) describes the accuracy of the time between the absorption of a photon and the dedicated evolution of a detector response. In most of SNSPD publications this parameter is also called detector jitter (t_j).

One has to divide the t_j in two parts: The intrinsic time jitter and the time jitter which originates from the amplitude jitter of the SNSPD response.
The intrinsic jitter can be seen as random delay between the moment of photon absorption and the occurrence of a resistance in the meander. The intrinsic mechanism in the detector that produces the uncertainty is still unclear. It might be depending on the hot-spot evolution. Depending on the grains in the material, the exactness of the etched width at the absorption position and the heat transport in the superconducting film (diffusivity D and electron-phonon interaction time τ_{e-ph}), the time of the resistance evolution might vary. In [29] a τ_{e-ph} of 7 ps is reported. A mean hot-spot growing time, which correlates to τ_{e-ph}, of 15 ps is presented in [30]. The absolute values of τ_{e-ph} do not define the jitter, but the deviations from these average values. One can at least assume that these deviations are at least shorter than the mean values. Indeed, the growing time is slightly longer or shorter, depending on the geometric conditions on the point of absorption. However, a strong enhancement of the time of the hot-spot growth based on geometric reasons would obviously influence the detection efficiency as well as and the cutoff frequency of the wavelength, which is not probable according to the results in section 4.1. To reach such deviation in time the dedicated geometric inhomogeneities should be in a range where they are visible by SEM imaging, which can not be found for all samples in this thesis. Therefore, one can assume an intrinsic jitter of an SNSPD in the range of $\ll$ 15 ps, although this is only a rough estimation. Further analysis of the exact definition of the intrinsic jitter and possible improvements for its reduction are still open questions.

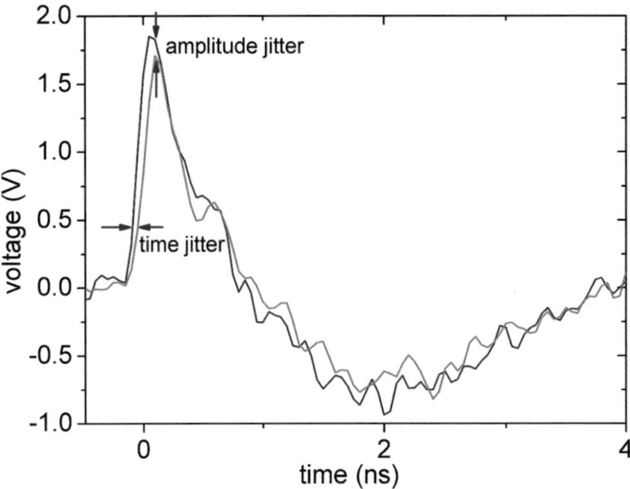

Figure 2.9: Time and amplitude jitter of an SNSPD. The two pulse responses originate from two single-shot measurements by a 6 GHz oscilloscope.

The amplitude jitter [52] is related to the readout impedance. The response pulse originates in a current redistribution in the detector path. This redistribution is a dynamic process, which means, the current will be redistributed and return after a certain time. These times are defined by the components in the readout circuit and the impedance of the detector itself. The detector impedance is a dynamical value, as well. In the stationary case, the detector is purely inductive. With the evolution of the normalconducting belt and the growth of the belt walls, this impedance gets an additional time-dependent, resistive component. In a dc consideration, this would lead to a dynamic time constant $\tau = \frac{L_{det}(t)}{R_{det}(t)}$ (also depending on local material and geometry parameters and absorption conversion factor), which individually varies the amplitude of the response pulse. In Fig. 2.9 one can see that the amplitude jitter directly leads to an additional time jitter t_j. In an rf consideration the matching between the changing source impedance and the 50 Ω readout impedance varies during the lifetime of the detector pulse, which also individually affects the pulse shape.

Several groups present measured values of t_j between 18 ps and 35 ps [16], [53], [54], [TGS$^+$12], however, it is very difficult to correlate these values to significant geometry, material or operation parameters. In [55] it is discussed that the t_j depends on the spatial position (edge or center of the active detector area). Other publications describe that the detector performance generally suffers under the constrictions due to the meander turns [56], [49], which could influence the t_j, as well.

Furthermore, the readout bandwidth defines the steepness of the slope of the SNSPD pulses, which correlates with the amplitude jitter and therefore, with t_j, as well. The bandwidth dependent measurements in [16] confirm this dependence.

2.2.5 Multipixel and multiphoton detection concepts

The SNSPD research is mostly concentrated on single-pixel detectors because the optimization of any performance parameter can always be tested with a single-pixel detector. However, from the application's point of view it would be quite convenient to get a multipixel system. This would allow true imaging applications or it would allow scanning systems, which enable fast 2D-sampling of molecules as used in spectroscopic research [57]. A multipixel concept can also increase the system detection efficiency because the active detector area can be increased using several pixels.

The SNSPD responses are measured in the time domain, which requires new multiplexing concepts. Up to now, first multipixel concepts exist, e.g. using several separate readout channels [43]. In [58] a further concept is described, which is based on the pulse amplitude variation caused by different numbers of temporally parallel photon events. This readout could be even used for multiphoton resolution. Multipixel concepts based on RSFQ electronics as cryogenic readout electronics [59], [60], [OHF$^+$11] are more and more in the focus of research, as well.

One can determine that the resolution of multiphoton events and multipixel detection is a related task and no multipixel readout concept is able to cover all types of applications. Individual strategies are required to solve individual multipixel tasks.

2.3 Summary

The SNSPD pulse is the response of the detector to the absorbed energy of a single photon. The intrinsic scattering processes, described by the two-temperature model, leads to a reduction of the superconducting order parameter and a hot spot in the nanowire. If the hot spot reaches a certain strength, it results in a temporally short evolution of an normalconducting belt in the current carrying narrow and thin nanowire. The relaxation of the detector response is related to the thermal conditions in the nanowire and the embedding readout circuit, described by an electro-thermal model. In this model, the kinetic inductance, a type of inductance which dominates only in superconductors, strongly influences the time scale of the evolution and relaxation of the detector response.

The characteristic parameters of an SNSPD are the detection efficiency, which typically reaches values up to 20% and with improved absorption concepts even values close to 100%, the dark count rate, which lies depending on the operation condition at values $< 100 \text{ s}^{-1}$ and the time jitter, which reaches values below 20 ps. The unique performance parameters and the adaptable design of an SNSPD allow interesting applications e.g. in the field of spectroscopy. However, the development of multipixel chips constitutes a challenge in the SNSPD development.

3 Cryogenic setup for characterization of SNSPDs

The following chapter describes the self developed IMS dip stick cryostat in detail, which is used for several analysis task [RPH$^+$10], [HDH$^+$12], [HRH$^+$12], [HWE$^+$12], [HRI$^+$12], [HAI$^+$13], [RHH$^+$13]. Any deviance from the description here, which refer only to one of the experiments in the thesis, are separately described in the section of this experiment.

3.1 Dip stick cryostat

Superconducting nanowire single-photon detectors require cryogenic conditions for operation [61] because the critical temperature for NbN detectors is $T_C < 15$K. Depending on the operation condition and the kind of radiation coupling, there exist different types of cryostat systems [62].

At IMS it is possible to develop and fabricate SNSPDs completely in house. Therefore, the number of tests of detectors can be quite high. For such measurements a dip stick cryostat is more suitable than a bath cryostat. A dip stick is a tube construction, where the cryogenic sample is moved into the dewar to the cryogenic fluid. A dip stick system allows fast cooling, pumping and warming up. Even several detector samples can be prepared on individual carrier blocks for fast exchange. Concerning the expected experiments and physical investigation, the dip stick was optimized for the measurement of several detector parameters. Additionally to the measurement of a high accurate detection efficiency in a broad spectral range, the system is able to measure dark counts and a fast time jitter (t_j) of the detectors. Moreover, the system is able to operate at different temperatures and allows flexible implementation of additional electronics components.

The cryogenic measurement setup as outlined in Fig. 3.1 consists of the dip stick cryostat with its internal components, but contains also several additional components at room temperature, which are necessary for operation and data acquisition. The individual components are described in the following section.

The cryostat (see Fig. 3.2) consists of a tube construction, which is moved into a cryogenic dewar during operation. The bottom part is in contact with the cryogenic liquid and the top part is at room-temperature (see Fig. 3.2, a)). The stick can be cooled down e.g. to 4.2 K by liquid helium. The detector samples and internal electronics are not in direct contact with the liquid helium but separated by an outer stainless steel tube. The bottom part of this outer tube around the detector consists of copper. This two-metal construction

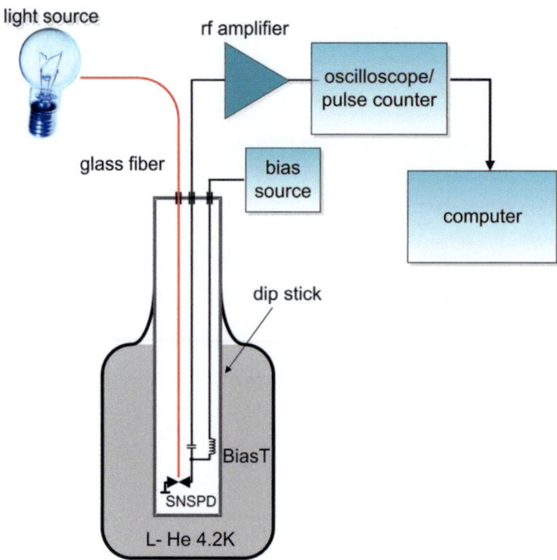

Figure 3.1: Simplified scheme of a cryogenic measurement setup based on a dip stick cryostat.

allows a good thermal contact to the liquid at the position of the detectors, but a worse thermal path in direction to room temperature (see Fig. 3.2, a) left). The inner part of the tube can be pumped down to allow operation at vacuum condition down to a pressure of $< 10^{-3}$ mBar. The vacuum pump is connected to the top of the cryostat by a small vacuum tube (see Fig. 3.2, b)).

The cooling of the detectors is realized by two methods. First, it is possible to fill gaseous helium into the tube by a needle valve (see Fig. 3.2, b), black valve on the right side). This improves the thermal conductivity to the cryogenic outer tube. The gas is directly taken from the dewar by a thin flexible tube and therefore, has almost pure helium content (the flexible tube is not shown on the image). The amount of gas can be measured indirectly by measuring the pressure in the tube at the top of the dip stick (see Fig. 3.2, a), red pressure meter with cable). It can be varied during the measurements if required. Moreover, there is a thermal contact between the inner construction with the mounted detector and the outer tube close to the detector position by two rings of Copper-Beryllium springs, which are soldered to the detector block (one of them can be seen quite good on Fig. 3.2, e)). If the pressure is low, the detector can be operated even at higher temperatures than the cryogenic liquid temperature, in connection with a heater (resistance in the detector block). This allows measurements from 4.2 K up to the transition temperature T_C of the detector sample or even higher, if necessary, which can be controlled by a temperature sensor in the detector block. Since the detector is mounted on a quite massive copper block the detectors can be operated at stable temperatures. Several experiments have shown that the presence of the contact gas does not influence the detector performance as long as the temperature is stable. The inner construction, where the detector is mounted on the end, consists of a stainless steel skeleton (see Fig. 3.2, a) right). This provides good stability, little thermal load and protection of all connection lines. Moreover, it delivers space for additional electronics like filters and cryogenic amplifiers. All signal and power supply lines have a vacuum feed trough on the top of the stick to contact them with room temperature electronics (see Fig. 3.2, c)).

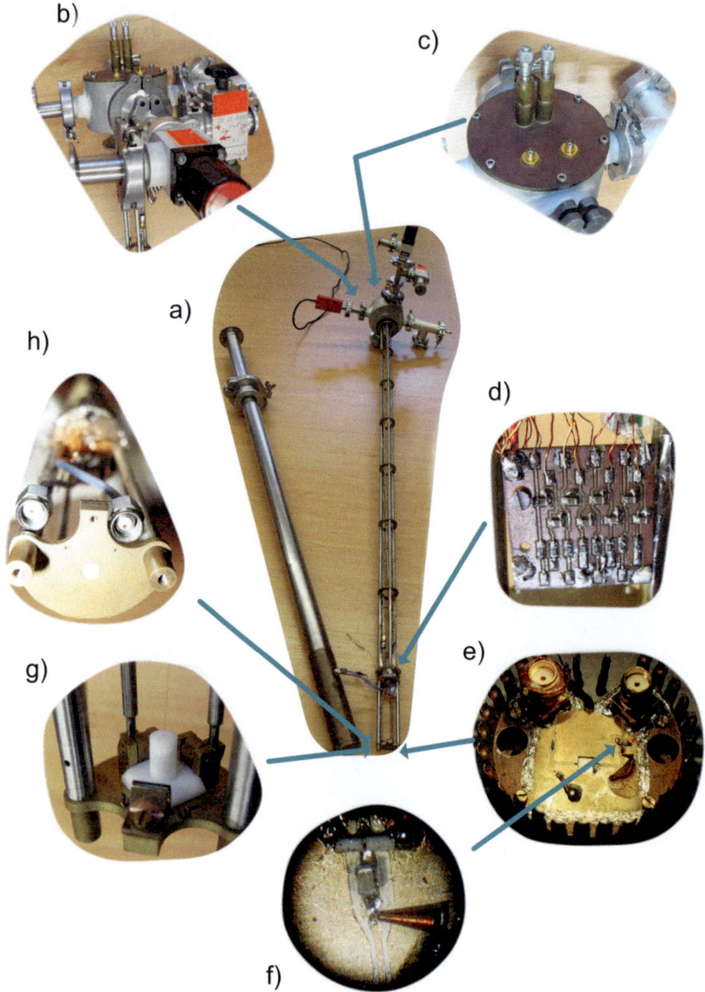

Figure 3.2: (a) Image of the whole dip stick construction. The stainless steel tube is assembled over the dip stick skeleton during operation. (b) Valves for pumping and contact gas control. (c) Vacuum feedtrough as connections for rf cables and positioners for fiber adjustment. (d) 1st order passive filter stage. (e) Carrier board for the detector chip with embedded biasT and cavity for the detector sample. (f) Embedded BiasT. (g) Adjustable Teflon sled with inserted fiber. (h)View from the bottom side. The detector can be illuminated by the fiber in the white Teflon sled through the hole in the brass plate. Also the SMA cables can be seen which are connected to the carrier board during operation.

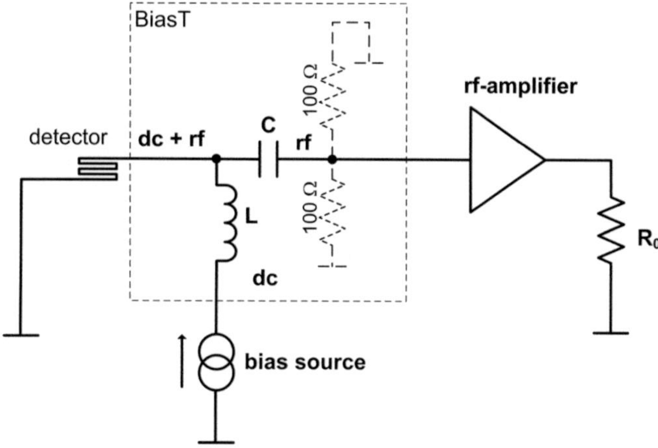

Figure 3.3: Scheme of a biasT. The dashed components are mounted for better matching, however, only in some of the later experiments.

3.2 BiasT design based on hybrid components

All SNSPDs work as an active element. That means a certain power supply is required, to provide the bias current I_B. Superconducting detectors can be operated either with a constant current biasing or a constant voltage biasing scheme. However, the voltage signal of SNSPD is a rf signal. The dedicated bandwidth starts from several MHz and goes up to several GHz. The detector response evolves by a fast redistributing of the bias current from the detector path in the readout path. That means, the dc bias current and the rf signal, are not galvanically separated. However, the readout electronics requires the pure rf signal, which therefore must be separated from the dc signal.

The splitting of these two signal types is realized by a so-called biasT. The biasT operates as frequency-separating filter stage. It can be mounted close to the detector at cryogenic conditions or at room temperature in front of the rf amplifiers. The advantage of the localization on cryogenic stage is that the noise, which occurs on the bias line, can be filtered out before it reaches the rf line, which improves the signal-to-noise ratio (SNR) of the signals in the readout line.

The biasT consists mainly of two electric components. A capacitance C in the rf path to prevent the dc signal going along the rf path and an inductance L which hinders the rf signal vanishing along the dc path. The basic scheme is outlined in Fig. 3.3.

The frequency range of a typical SNSPD response is quite depending on the design but roughly lies between several MHz and several GHz (see Fig. 3.4). Based on the experience of [63], which focuses on a comparable frequency range, it has been estimated that a coil inductance of 840 nH at room temperature (series resistance of the inductance $< 0.02\ \Omega$ at 4.2K) and a capacitance of 12 nF at room temperature are suitable components to guarantee

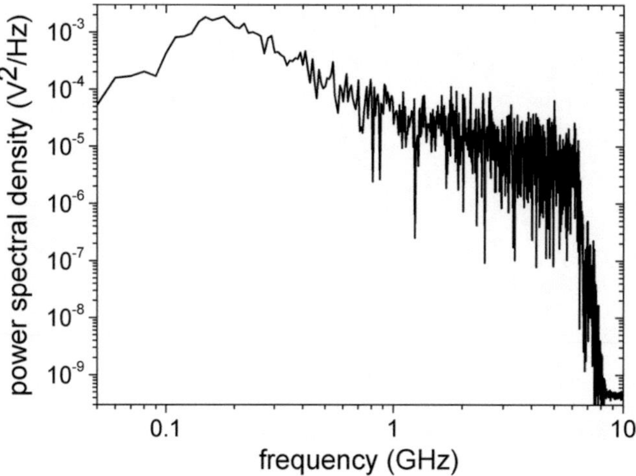

Figure 3.4: Power spectral density of an SNSPD pulse from a 4 nm thick detector biased with 17 µA. It bases on a FFT of the data of a 6 GHz oscilloscope.

a frequency response in the transmission path covering the required frequency range of SNSPD signals.

The biasT components have to be embedded in a microwave line for reflection-free transmission. The line impedance should be matched to the commercial readout components and is therefore $R_0 = 50\ \Omega$. For good contacts between biasT and detector sample, the same type of planar waveguide is selected as on the detector chips itself (see section 4.1.2), which is a coplanar structure. This allows direct bonding of both, the inner transmission lines and the GND planes with little rf mismatch. The waveguide is made from a RogersTMM10i substrate with 35 µm copper layer on both sides. The biasT components are soldered on the coplanar structures. The coplanar waveguide is defined by several geometric parameters, which are depicted in Fig. 3.5. Several relations between g, w and h enable a line impedance $R_0 = 50\ \Omega$ because only the relation of these parameters is essential. Suitable values are simulated by the simulation tool Microwave Office, TXline. Here, two different sets of parameters g and w of the coplanar structure (h is fix) are selected to keep, on the one side, the transmission line small but to enable partly a broader transmission line to integrate the soldering pads of the hybrid biasT components, as well.

The biasT is characterized with a test sample consisting of a 10 mm long coplanar line and with all biasT components soldered on it, which enables a 2-port measurement with the network analyzer. The biasT sample is embedded in a brass holder and connected by SMA housing connectors. A 2-channel rf dip stick is used for cryogenic S-parameter measurements. The measurements are done by a 67 GHz network analyzer from Agilent.
Figure 3.6 shows the S-parameter measurement of the biasT at room temperature and at 4.2 K. The transmission $|\underline{S}_{21}|$ shows in both cases up to 8 GHz an almost flat course. The

reflection $|\underline{S}_{11}|$ is lower than -10 dB in the presented frequency range. The cooling of the dip stick also cools the rf lines, which leads to lower attenuation of the rf lines with decreasing temperature. Therefore, the cryogenic $|\underline{S}_{21}|$ parameter is even slightly better than the room temperature $|\underline{S}_{21}|$ parameter. Although the required bandwidth of an SNSPD pulse is geometry-dependent and therefore variable, 8 GHz is a sufficient upper limit. The $|\underline{S}_{11}|$ parameter increases close to 30 MHz. This indicates that the biasT capacitance starts to constrict the transmission of the signal, however, the damping of $|\underline{S}_{21}|$ is still small. Since the readout is also limited by the amplifier bandwidth, which has the lower frequency limit between 10 MHz and 50 MHz (see section 3.3) the biasT covers the required frequency range.

The PCB design of the final carrier board is shown in Fig. 3.7. The carrier board allows to readout two channels (two coplanar lines) or can be used for two port transmission measurements. The detector is mounted in the square in the middle of the carrier board and bonded to the coplanar waveguide. The capacitance is mounted on the gap in the transmission line. The inductance is soldered to the inner lead of the transmission line between capacitance and detector at the solder contact of the capacitance. The other side of the inductance is soldered to a wire, which is guided out to room temperature. The carrier board is fabricated by a standard photo-lithography process, based on a HR mask (high-resolution photo-lithographic mask) and etched in a PCB wet-etching process. The HR mask allows to produce geometric structures with a resolution of less than 10μm. On top of the copper layer a thin gold layer is added by a galvanisation process (the small copper bridges on the mask are required for this process).

An image of the soldered biasT construction can be found in Fig. 3.2 e) + f). In front of the biasT, two 100 Ω resistances are added in parallel to improve the 50 Ω matching of the line (see dashed components in Fig. 3.3). These 100 Ω resistances are mounted only in the TTS measurements.

For typical SNSPD readout only one line of the two channels is required, however, the complete setup is design with two complete rf paths, allowing transmission measurements, as well.

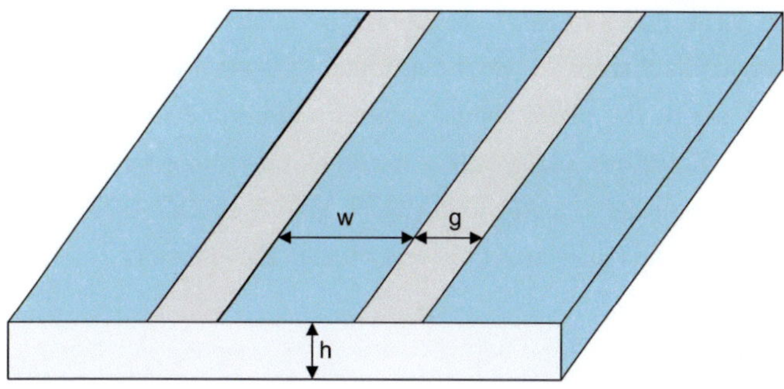

Figure 3.5: Scheme of the coplanar waveguide.

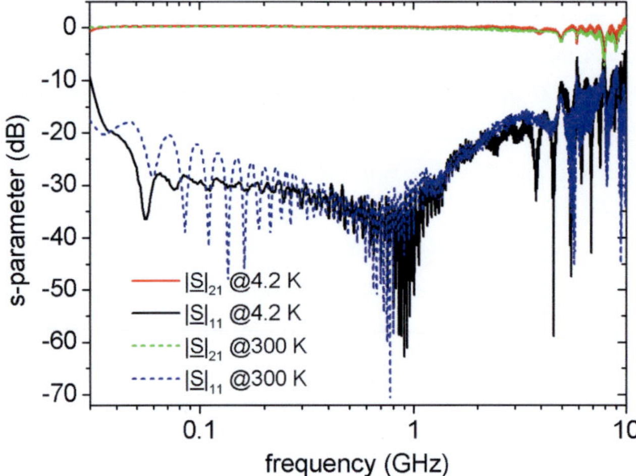

Figure 3.6: Measured S-parameter of the cryogenic biasT made with discrete elements and embedded in a coplanar waveguide.

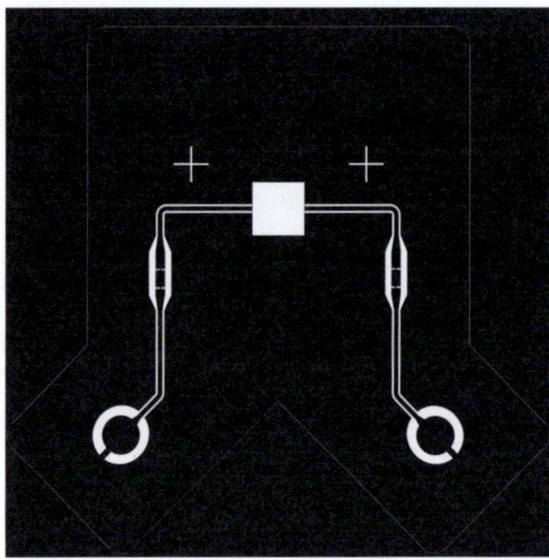

Figure 3.7: Mask of the two-channel carrier board including footprints of the biasT.

Although the performance of this hybrid biasT is suitable, it is not the final solution for all readout concepts. In chapter 5 several multipixel readouts are described. A space-saving routing and mapping of the pixels would require a direct biasing close to the pixels for individual biasing. Therefore, such readout would gain in future by a cryogenic integrated planar biasT directly on the detector chip, which allows an increased density of detector pixels. The feasibility of such a concept is described in the appendix.

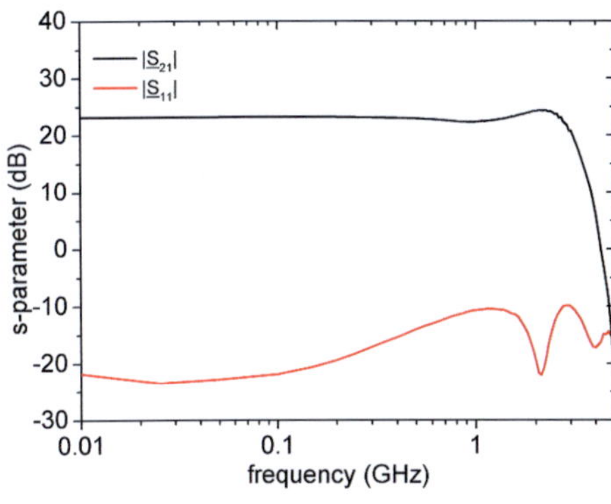

Figure 3.8: S-parameter of the type 1 (ZFL-2500VH+) amplifier.

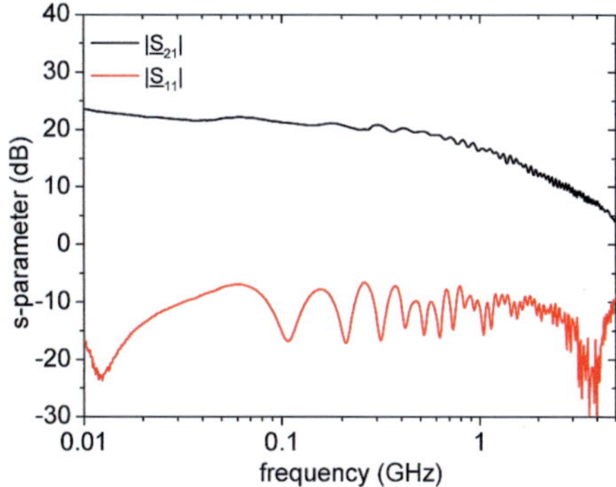

Figure 3.9: S-parameter of the type 2 (ZX60-33LN+) amplifier.

3.3 RF readout of SNSPDs and data acquisition

The rf signal is guided out by a semi-rigid cable. On the top of the dip stick, there are two rf-line vacuum feed troughs. The cable and connector type of the whole readout is SMA, which guarantees an rf transmission up to 26 GHz, which is far above the required frequency range for SNSPD readout.

The complete process of amplification and data acquisition is performed at room temperature. For amplification two types of rf amplifiers from Minicircuit are used. Type 1 delivers a gain of 23 dB and a bandwidth from 10 MHz to 2.5 GHz (see Fig. 3.8). Type 2 delivers a gain of 23 dB and a bandwidth from 50 MHz to 3 GHz (see Fig. 3.9). Depending on the bias current I_B of the detector, the response amplitude might be to low for readout with one single amplifier. Consequently, a cascading of several detectors might be necessary. The noise figure of the amplifiers in the readout chain is $NF \approx 1.1$dB (type 2) and $NF \approx 5$dB (type 1). If more than one amplifier is required and still the bias currents I_B are in the range, where its dedicated responses are critical to resolve from the noise floor, the type 2 should be used as first amplifier in the chain because the noise figure of the first component in the chain defines mainly the common noise figure of the chain ($F_{tot} = F_1 + \frac{F_2}{G_1-1} + ...$). A typical configuration is shown in Fig. 3.10. If a more flat frequency transmission is required, the type 2 should be removed, since the type 2 delivers the worse $|\underline{S}_{21}|$ (see Fig. 3.9) compared to the type 1 (see Fig. 3.8)

Generally, lower noise figures would further decrease the noise floor of the readout, however, all detectors with a critical current $I_C > 10\mu$A deliver in this setup a sufficient pulse amplitude to enable counting with our counter electronics.

Although cascading of several amplifiers increases the overall gain, there are disadvantages. An amplifier gain is only available in a certain frequency bandwidth limited by its upper and lower 3 dB cutoff. A cascaded order of several amplifiers reduces the effective bandwidth of the amplifier chain because the 3 dB cutoff of each amplifier in the chain leads

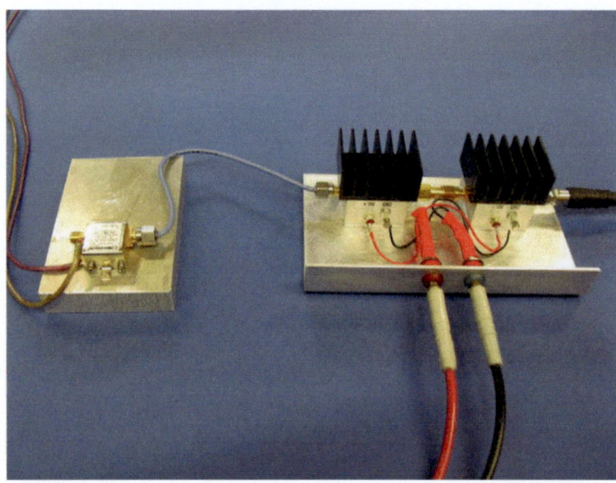

Figure 3.10: Cascaded rf amplifiers typically used in SNSPD measurements.

to a higher attenuation, which behaves like a shift of the 3 dB cutoff to lower frequencies. The effective upper cutoff frequency of an amplifier chain can be calculated [11]:

$$f_{eff} = \left[\sum_i f_i^{-2} \right]^{-\frac{1}{2}} \tag{3.1}$$

f_i is the cutoff frequency of the i-th amplifier in the chain. According to Eq. 3.1, the chain of the three amplifiers in Fig. 3.10 has an effective upper bandwidth limit of 1.7 GHz. Finally, one has to note that the same effect is also given for the effective lower 3 dB cutoff. A suitable f_{eff} is defined by the detector and the type of measurement, e.g. the SNSPD pulse bandwidth is strongly dependent on the detector geometry, since it defines the critical current and the inductance of the detector wire. As long as only the pulse events need to be counted, even a lower f_{eff} of several hundreds of MHz is sufficient. On the other side, it was shown that for jitter measurements a high f_{eff} improves the time resolution [16]. Hence, for detailed analysis of pulse amplitude, the bandwidth f_{eff} should be enlarged as much as possible by both increasing the upper limit and decreasing the lower limit of the system band gap. Otherwise, the response pulse is mainly defined by the electronics and not by the detector.

After amplification the detector signal has to be recorded for data analysis. Suitable acquisition electronics should enable count measurements, analysis of amplitude and time resolved measurements and cover the bandwidth of the SNSPDs. In all experiments either a 500 MHz, a 2 GHz, a 6 GHz or a 32 GHz real time oscilloscope, or the Stanford & Research SR620 pulse counter (Bandwidth 300 MHz) is used. For TCSPC measurements, a TCSPC electronic from Becker & Hickel (Simple-Tau-130-EM-DX) is used in combination with the analog front end (AFE) described in section 3.7.2.

3.4 Low noise setup and current source for detector bias

Each active and passive component in a electronic system is a certain noise source or adds distortion to the signals or supply currents, respectively. In opposite to hot-electron-bolometers (HEB) or transition edge sensors (TES), it is not possible to reduce the noise impact by averaging an SNSPD signal. Typical bias currents of SNSPDs are located in the range of several tens of µA, e.g., a 4 nm thick NbN detector with a nominal width of 100 nm reaches values down to 17 µA [HRI$^+$10b]. As soon as the superposition of the bias current I_B and the additional distortion or noise peak produces a total current in the range of the critical current I_C of the detector, the detector becomes normal conducting at least as long as I_B is not reset to zero (so-called detector latching). That means, noisy and distorted systems will bring the detector either in a continuous latching state or reduce the performance of the measurement due to the necessity of operation at lower I_B.

The origin of the noise and distortion is quite variable. Typical noise sources are already part of the bias electronics or are coupled in the electronic system. All electronic components in the bias electronics have a thermal noise v_{therm} and the typical noise sources of semiconductors, like 1/f noise, popcorn noise, shot noise.

As long as the noise source is localized at a defined point in the setup, one can try to filter it out. However, the thermal noise v_{therm} is generated in the full setup, where a certain resistance R exist. Its sources are the detector leads, the filters and the lines in the cryostat. The thermal noise v_{therm} is defined by:

$$v_{therm} = \sqrt{4 \cdot k_B \cdot T \cdot R \cdot \Delta f} \tag{3.2}$$

According to Eq. 3.2 v_{therm} is depending on temperature and uniformly distributed over the frequency spectrum.

One has to consider two ways of delivery of the noise to the detector stage. The first is the bias line and its ground (GND) lines. Here, a cryogenic filter in the bias line is quite suitable to reduce at least a part of v_{therm} and other noise influences close to the detector position. Therefore, several filter stages in the dc path limit the bandwidth to a maximum frequency of about 1 kHz. Some of the filter stages are inserted at room temperature, and one filter is realized as a cryogenic passive RC filter stage for all dc lines going to each cryogenic circuit (see Fig. 3.2 d). Theoretically, the bandwidth can even be further reduced, but during operation the system needs even in the dc path a certain bandwidth for fast reset and bias current modifications. Hence, a filter can not remove the complete noise impact. Consequently, the selection of all components in the readout and supply electronics has to be done in consideration of low noise performance.

The second noise delivery to the detector is the rf readout path, which definitely requires a broader bandwidth than the dc path. In the rf path, it is possible to add a dc blocker, which sets the lower cutoff frequency to the range of the rf amplifiers (10 MHz). This blockers exist in two versions. Either, only the inner conductor is interrupted for dc-signals by a capacitance, or both inner and outer conductor are interrupted (Both types also reduce the thermal conductivity of the coaxial cable and therefore improve cooling.) Moreover, the amplifiers should have a low noise generation, and the power supply of these amplifiers, as well.

The rf line can have further distortion sources. The SNSPD is not a source with 50 Ω impedance, which leads to a mismatch between the detector and the readout/amplifier impedance. This mismatch leads to reflections, which are guided back to the detector stage. The reflections can enhance the dark count rate. To reduce this influence to an uncritical level, a rf attenuator (3 dB or 6 dB) can be mounted in the rf line between the detector line and the amplifier input. This method is used in part in the later experiments.

Another important design parameter is the correct guiding of the GND currents. The complete dip stick is made from stainless steel or copper, which behaves as electrical conductor line. The detector holder and the rf carrier board are galvanically connected to the dip stick. Therefore, the detector GND is at the same potential level as the dip stick construction. The use of different materials in the GND reference, with different specific resistances may lead to a potential drift along the setup, which can also be enhanced by the Seebeck-effect [64]. Moreover, the existence of different routes for the backward dc current can produce GND loops, which behave as a noise coupling antenna. Therefore, a bad GND guiding can produce distortion peaks both on the readout screen and on the detector itself, which influence the performance. Consequently, the complete setup is designed to have only one defined GND path between bias source and detector. Moreover, possible GND loops in the setup are interrupted. Nevertheless, all metal parts, which belong to the GND level, are connected at one point in the setup (this includes the bias source and shielding GND, as well) to have no floating components in the system. The already mentioned double-path dc blockers are even for this task quite helpful because it interrupts the dc GND connection of the rf line, which is in parallel to the stainless steel tube and the GND bias line.

The selection of low noise bias source is quite important. Several commercial sources were tried, but no source was able to cover the range of requirements, which occur in the experiments at IMS. Therefore, a self made bias source is developed in co-work with Matthias Arndt and Petra Thoma [65]. The **requirements** of a bias source for SNSPD readout are:

- low intrinsic noise $i_{pp} \ll 1$ µA,

- manual operation with step resolution of current of ≈ 0.1 µA,

- battery supply,

- selectable current values between 0 µA and 200 µA,

- long time stability of several hours.

The developed bias source is designed to be operated in current- and voltage-controlled mode. The driver component of the bias source is an ultra-low noise operational amplifier (LT1028) trimmed to a system bandwidth below 1kHz. The source is controlled by a potentiometer allowing a current resolution of 0.1 µA. The power supply is given by a battery pack of ± 10 V. To save battery power the circuit is optimized for low quiescent currents, which guarantees measurements with long time stability of several days. Except the displays, the complete circuit is realized by a pure analog design, and therefore free of any clocked distortion. The peak-peak voltage noise at 1kHz of this source is $v_{pp} < 2$ µA, leading to a current noise of $i_{pp} < 400$ pA at 1kHz with the internal resistance of the source. An image of the final source is shown in Fig. 3.11.

Figure 3.11: Image of the low noise current source

3.5 Fiber-based radiation coupling

The position of the detector is far inside the helium dewar. A direct free-space-coupling is not possible for this reason. Instead, a glass fiber is used. Fibers allow a flexible guiding of radiation, even at vacuum conditions and in cryogenic areas. Moreover, a plug-and-play connection of the cryogenic setup to the optical application is possible [TGS[+]12],[66]. The basics about fiber optics can be found in standard optics literature [36]. The discussion in this section is only focused on the specific challenges of SNSPD coupling.

One task of the dip stick is the measurement of the detection efficiency. For this measurement, an accurate estimation of the absolute number of photons on the active area of the detector is required.

Single-mode fibers are suitable only for a very narrow spectrum (few 100 nm), which on the

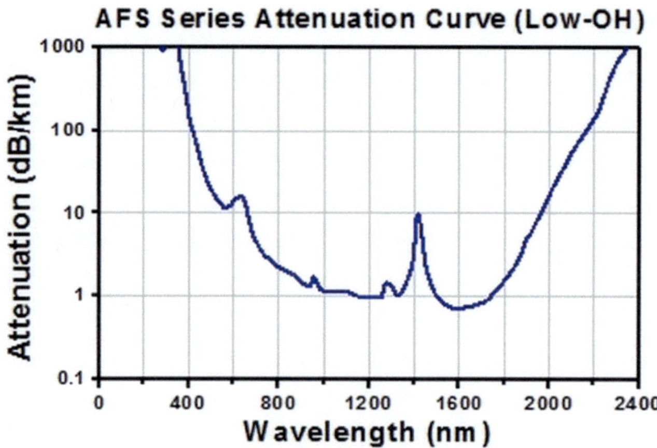

Figure 3.12: Attenuation profile of the step index multimode fiber used in the dip stick cryostat [70].

other side is enough for applications like spectroscopy [TGS$^+$12] or quantum communication [3]. The cross-section of the fiber core, which guides the largest amount of radiation, fits well to the typical detector sizes (44 μm^2 to $15 \cdot 15 \ \mu m^2$). The detectors can be optimized to the fiber geometry [67] and there is a quite suitable method to couple the fiber end and the detector with low loss using fix mounted fiber ferules on the top of the detectors [68]. Moreover, a better coupling can be reached by fabricating the detector directly on the fiber tip [69].

If the operation in a wider spectral range (optical to infrared) is required one has to use multimode fibers. However, these fibers have a core diameter of 50 μm and larger, which increases the coupling losses between fiber end and detector and leads to an *OCE* « 100% . If the *OCE* can be estimated, a multimode fiber is a suitable fiber to define the detection efficiency in a wide spectral range. Therefore, the AFS105/125Y is used in the cryogenic dip stick which covers a frequency range from 400 nm to more than 2 μm (see Fig. 3.12). The fiber end is cut by a fiber cleaver, which leads to a surface roughness of the fiber front comparable to a polished fiber plug. The fiber end is mounted in a Teflon sled, which can be moved in x/y direction by positioning screws (see Fig. 3.2 g)). The maximum traveling range of the x/y positioning stage is in the range of 1 mm.

3.6 Radiation sources and optical source calibration

Radiation sources

The experiments in this thesis are made with different radiation sources. For the spectral characterization of an SNSPD, a white halogen light is used. The light is spectrally split by a grating based monochromator (DON-DM151i from Zolix). The monochromator has two gratings: grating 1: 1200 L/mm, 500 nm blaze; grating 2: 600 L/mm, 1 μm blaze. The monochromator covers the spectral range between 400 nm and 2000 μm and is computer controlled by Labview.

All pulsed measurements are realized by a fiber based fs-laser from Menlo Systems. The system parameters are: Er-doped fiber oscillator; repetition rate 100 MHz $+/-1$ MHz, center wavelength 1560 nm $+/-20$ nm, linearly polarized, average output power > 30 mW, pulse length < 150 fs; FC/APC fiber connector. Since the laser source is based on fiber, it is optimal for fiber coupled measurements.

Some of the measurements are made with a 650 nm laser source. The source is a DFB laser diode, which can be operated either in cw mode or can be modulated with frequencies below 100 Hz. The modulation can be applied by directly modulating the bias current of the laser diode. In principle the output reaches up to 40 mW, however, only a part of the radiation is coupled into the fiber.

Calibration

The measurement of the spectral detection efficiency requires the knowledge of the photon rate n_{photon} crossing the active area of the detector, which is $4 \times 4 \mu m^2$ in most of the experiments. To define the correct photon rate $n_{photon}(\lambda) = P_{in}(\lambda) \cdot \alpha(\lambda) \cdot hc/\lambda$ on the detector area, one has to define all attenuation and coupling losses $\alpha(\lambda)$ in the optical system between radiation source and detector area.

First one has to define $P_{in}(\lambda)$. All available light sources are either fiber based, or allow a light collection and coupling into the fiber. The reference point of the incoming optical power is the input feed through of the cryostat. At this point the exact optical power $P_{in}(\lambda)$ per wavelength can be measured by a radiation power meter.

The second step is defining the optical power $P_{detector}(\lambda)$ on the detector to estimate the loss factor $\alpha(\lambda) = \frac{P_{detector}(\lambda)}{P_{in}(\lambda)}$ and finally to estimate $n_{photon}(\lambda)$. For this reason knowledge about the light guiding in the cryostat is required. The fiber outside of the cryostat is connected by two fiber SMA couplers to the fiber vacuum feed-trough and the internal fiber. The final end of the fiber at the cryogenic stage is cleaved with a $5°$ angle to prevent reflection in the fiber path and is mounted in the mechanical positioning sled. Due to the core diameter of the fiber ($D_{fibre} = 105$ μm) and the aperture of the fiber ($a = 0.22$), the fiber produces a light spot on the detector chip with a certain diameter $D_{spot} = f(D_{fibre}, a, d_{f-d})$ with d_{f-d} the distance between fiber end and detector. A high accurate measurement of the detection efficiency requires a homogenous $n_{photon}(\lambda)$ on the active area of the detector. Assuming that the end of the fiber is a point source, the light spot of the fiber is Gaussian distributed. However, this distribution can only be reached under the condition that only one radiation mode is guided in the fiber like in a single-mode fiber. In a multimode fiber, several modes can be guided, whose interference pattern depends on the angle of coupling, the optical power, and the fiber bends. Therefore, using a coherent light source, the output spot of a multimode

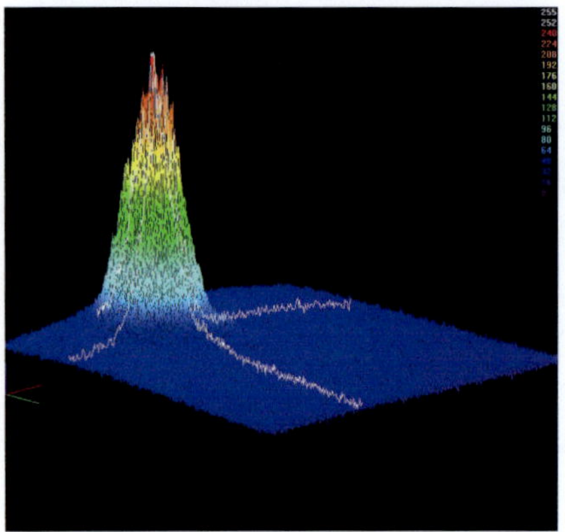

Figure 3.13: 3D beam profile of the fiber spot at $\lambda = 650$ nm coherent laser light

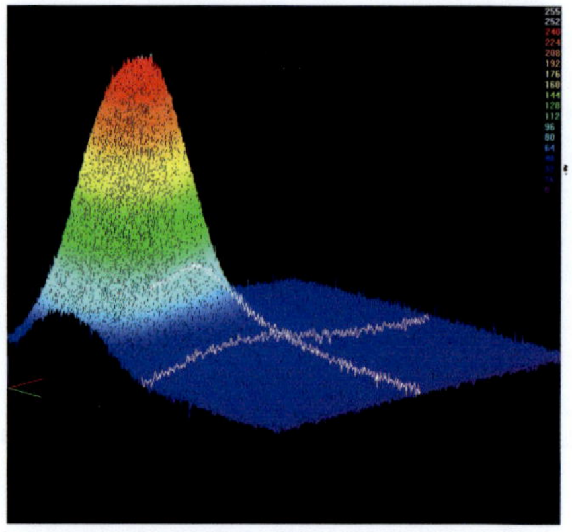

Figure 3.14: 3D beam profile of the fiber spot at $\lambda = 750$ nm non-coherent light.

fiber is typically Gaussian distributed but overlapped by a mixture of mode interferences, which lead to variation in the photon intensity on the spot (see Fig. 3.13). This intensity distribution varies even by moving the fiber, or by changing the mounting and light coupling and makes a predictable photon distribution impossible. The evolution of the modes is mainly depending on the coherency of the guided radiation. Non-coherent light does not have such a strong effect to a mode distributed spot because the interference behavior is due to the non-coherency of the light much less distinct. Since D_{fibre} is much larger than the detector size in case of a multimode fiber, the fiber tip is not a point source anymore. The spot of the fiber using non-coherent light is a Gaussian profile convoluted with the square function given by the core/cladding transition of the fiber (see Fig. 3.14), which broadens the plateau of the Gaussian peak.

The calibration requires a homogenous photon distribution on the detector. A Gaussian profile is not laterally homogeneous. However, if one enlarges the diameter of the Gauss shape by increasing d_{f-d}, the maximum of the Gauss profile also enlarges to an almost homogenous plateau (red and white area in Fig. 3.14). The mechanical positioning system of the dip stick is able to position the fiber end in a range of ± 0.5 mm in x/y direction over the detector area. If one selects $d_{f-d} > 1$ mm, then the radius of the $1/e$ decay of the beam spot has a size of > 0.5 mm, which allows to position a photon rate with a homogenous photon distribution $n_{photon}(\lambda)$ on the active area of the detector.

The calibration process is described in the following and also described in [71] more in detail:

Fig. 3.15 shows the scheme of the calibration procedure. The optical power of the monochromator output has to be defined for each required wavelength λ with a power meter at the end of the fiber in front of the fiber feed-through (reference layer 1 in Fig. 3.15). The grating in the monochromator splits the incoming light into its spectral parts, however, higher optical harmonics can overlay the basic wavelength, e.g. the wavelength range $\lambda > 800$ nm contains photons with $\lambda > 400$ nm. Since the detection efficiency of an SNSPD is different in these two spectral ranges, optical edge-filters are required to filter out the higher harmonics. To cover the spectral range between $\lambda = 400$ nm and $\lambda = 2000$ nm, two filters with the cutoff wavelength $\lambda = 750$ nm and $\lambda = 1050$ nm are required, which are inserted in the optical path during the wavelength shift. The part of the optical losses in the fiber inside the dip stick cryostat $\alpha_{fiber}(\lambda)$ is defined as the ratio between $P_{in}(\lambda)$ and $P_{cry}(\lambda)$, which is the optical power $P_{cry}(\lambda)$ of the fiber spot defined with three photo diodes at the end of the fiber at the detector position (reference layer 2 in Fig. 3.15). The visible range $(450 - 750$ nm$)$ is covered by a Si diode, the infrared range $(750 - 1700$ nm$)$ by a Ge diode and the small UV part $(400 - 450$ nm$)$ can be defined by a GaP diode.

To define the photon distribution of the convoluted light spot a CCD camera chip is required, which images the lateral spot of the fiber beam around the position of the detector (see reference layer 2 in Fig. 3.15). Fig. 3.16) shows a typical recorded beam profile made by the CCD camera at $\lambda = 650$ nm. The plateau with homogenous optical power is marked with dashed lines. The data of the CCD chip are only relative values and have to be referenced with the absolute optical power measured before with the diodes.

The monochromator gratings polarize the light of the halogen lamp. The *DE* of an SNSPD meander depends on the polarization (see Fig. 2.5). Consequently, an exact measurement of the *DE* should either have full polarized light or full non-polarized light, to realize correct evaluation. The required multimode fiber doesn't maintain the polarization, which leads to

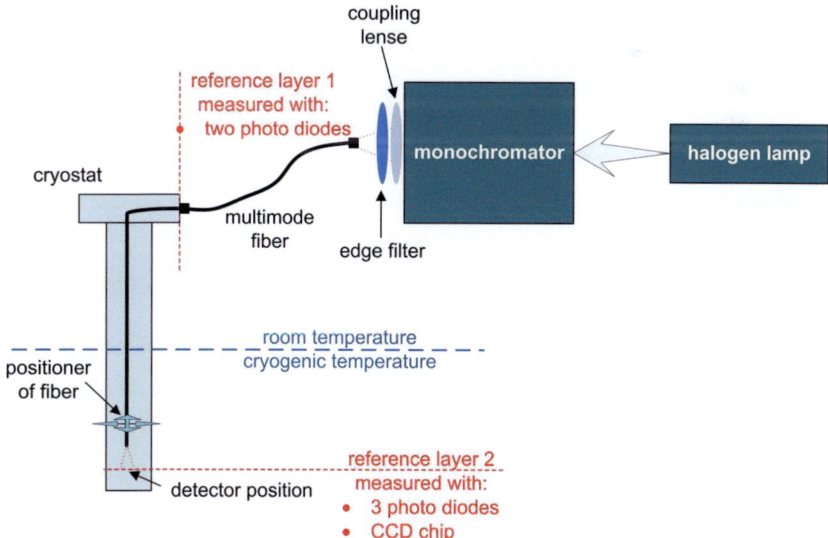

Figure 3.15: Scheme of the optical calibration of the SNSPD measurement setup for detection efficiency measurement

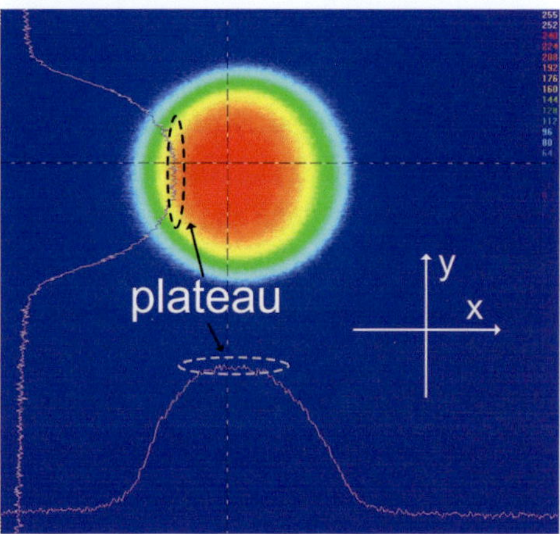

Figure 3.16: 2D beam profile of the CCD camera chip measured at $\lambda = 650$ nm. The plateau with homogeneous photon distribution is marked in the beam profile. The beam spot can be positioned in x and y direction on the chip layer.

a loss of polarization with fiber length. It could be found out that for a fiber length $l > 2$ m the remaining polarization is < 2 %, which is sufficient for a non-polarized measurement.

With the estimated $\alpha(\lambda)$ one can estimate n_{photon} depending on λ. The calibration reaches an accuracy of 4 % in this setup [72]. The factor $\alpha(\lambda)$ is in the order of 10000 and is valid as long as no modification on the setup is arranged. The only calibration parameter which has to be defined in each SNSPD measurement is P_{in}. A suitable power level for P_{in} is given by the requirements of the detector. In most cases, the detector has to be measured in the single-photon regime, which means that $n_{photon}(\lambda) << n_{max,det}$. $n_{max,det}$ is the maximum count rate of the detector. Moreover, the non-absorbed photons of the beam spot might lead to a heating of the copper block, where the detector is mounted, which changes the operation temperature. Therefore, P_{in} has to be weighted with the *OCE* to estimate the power dissipation around the local detector side, to reduce P_{in} to an acceptable level.

3.7 Readout concept for jitter measurements of SNSPDs with high time resolution

In the previous section the measurement setup was described concerning its general characteristics. Now a special measurement technique is introduced, the so-called time-correlated single-photon counting (TCSPC) measurement method. This method is important for time jitter measurements and it offers the possibility to reconstruct a very short radiation pulse by a statistical approach. In addition, a special analog front end for the TCSPC measurement is described which allows an analysis of the detector response with 10 GHz bandwidth.

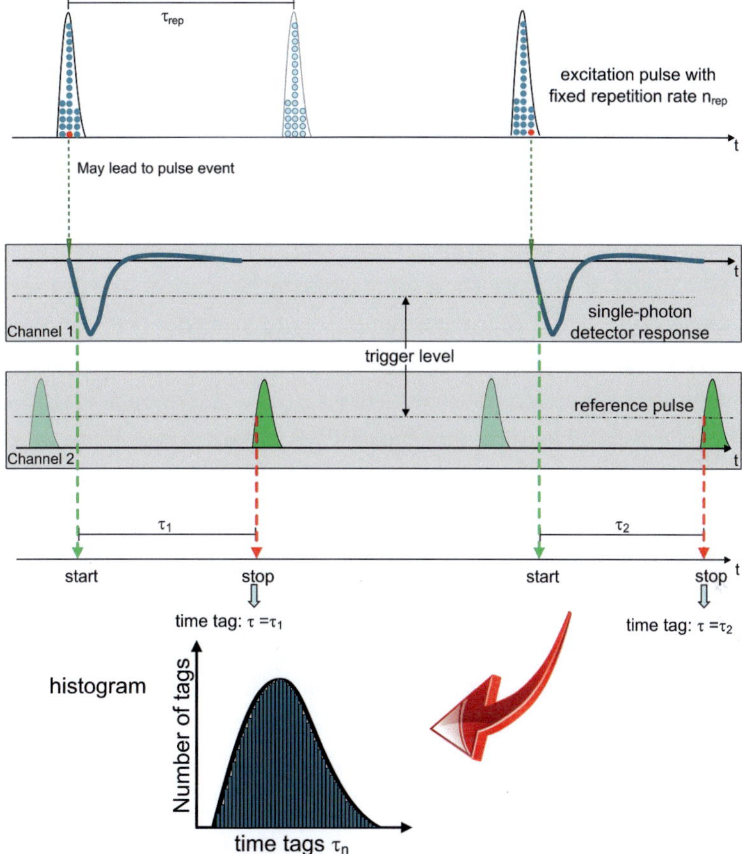

Figure 3.17: Principle of TCSPC. In the first time bar the original radiation pulse are depicted with its internal photon distribution. The two grey time bars show the pulse responses of the SNSPD and the reference signal which are connected to the input channels of the TCSPC electronics. The undermost graph shows the final histogram extracted from the measured and summed start-stop time intervals.

3.7.1 Time-correlated single-photon counting method

TCSPC is a method to characterize radiation pulses with high time resolution using a statistical approach [73]. The TCSPC technique requires for this measurement detectors which can resolve single photons as direct response. Furthermore, a reference signal of the pulsed radiation source is needed.

A radiation pulse has a certain amplitude shape. At the beginning of the pulse only few photons arrive at the detector. The pulse amplitude increases with time to a maximum, which leads to higher photon numbers on the detector, and decreases again. The photons of the front slope of the pulse arrive earlier on the detector than the photons of the decay of the pulse. After measuring the time between the detector pulse and the reference, one can use this time information (time-tag τ) to define from which temporal position the photon originated (see Fig. 3.17). Several measured time tags τ accord with their statistical distribution the distribution of the photons in the radiation pulse, if certain conditions are fulfilled: The measured count rate needs to be much lower than 1 per radiation pulse. If the count rate is higher then the probability that a photon event is registered by the detector, is already 1 before all photons completely hit the detector. That leads to a preference of the earlier photon events in the radiation pulse or even a partly disregard of the photons in the tail of the radiation slope. This effect is called the pile-up effect. Therefore, the system must work in the single-photon regime. The statistical distribution can be depicted by a histogram graph. The time resolution of the TCSPC measurement $t_{j,\,total}$ is shorter than the detector pulse width w_{det} because not the pulse width of the detector defines the time resolution, but the accuracy of the arrival in comparison to the reference pulse. This accuracy is defined by the timing jitter t_j (also called transient time spread TTS) of the detector. If one wants to measure the pure system jitter $t_{j,\,total}$ of the setup, the pulsed excitation should have a Dirac-like distribution. To resolve, in contrast, the pulse shape of the radiation source, the following inequation needs to be true.

$$t_{j,total} < w_{rad\ pulse} \tag{3.3}$$

This inequation postulates that the total jitter $t_{j,\,total}$ of the complete readout system has to be smaller than the time scale of the pulse under investigation. If this condition is not fulfilled, the reconstructed pulse in the histogram is broader than the original radiation pulse.

The $t_{j,\,total}$ contains all jitter influences $t_{j,\,readout}$ of the setup, like cabling, amplifier, TCSPC electronics and jitter of reference pulse. Depending on the selected electronics, values of less than 10 ps are possible [74]. A further parameter is the jitter $t_{j,\,pulsedlaser}$ of the pulsed laser, which should be smaller than the $w_{rad\ pulse}$, as well. All jitter parameters are uncorrelated to each other. Therefore, the system jitter $t_{j,\,total}$ is defined:

$$t_{j,\,total} = \sqrt{t_{j,\,det}^2 + t_{j,\,readout}^2 + t_{j,\,pulsed\ laser}^2} \tag{3.4}$$

Concerning detector development the crucial parameter in $t_{j,\,total}$ is the $t_{j,\,det}$.

As mentioned in section 2.2.4, an SNSPD is a suitable candidate for TCSPC measurements, due to its single-photon resolution [TGS⁺12], [55]. The typical scheme of a TCSPC measurement setup with SNSPDs is shown in Fig. 3.18.

Although $t_{j,det}$ is uncorrelated to $t_{j,\,readout}$, $t_{j,\,det}$ is dependent on the readout circuit, since the amplitude jitter depends on the readout conditions (see section 2.2.4). In principle, $t_{j,\,det}$ can be corrected from influences of the amplitude jitter. Zhao proposed in [75] a method derived from Bertolini and Coche [76]. This method enables to calculate the time jitter as consequence of an amplitude jitter by knowing the mean slew rate of the amplitude for each discriminator level. As long as the jitter is measured with a real-time oscilloscope this data can be additionally measured. TCSPC electronics can not be used for these ap-

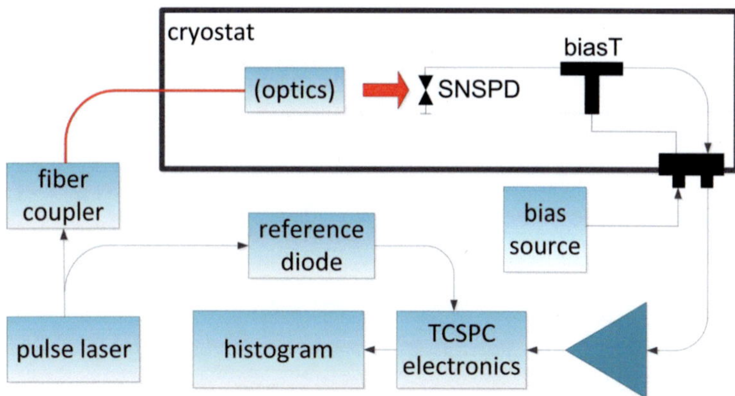

Figure 3.18: Scheme of TCSPC measurement with an SNSPD as single-photon detector

proaches because they can not measure the slew rate of the pulse. Instead, a typical TCSPC electronics uses a constant fraction discriminator (CFD) to reduce the amplitude jitter [77], which enables to optimize the amplitude jitter in a certain range of operation parameter. A TCSPC electronics like the "Becker & Hickel Simple-tau-130-EM-DX" has two parameters controlling the CFD. One is the CFD level and one is the zero-crossing (ZC) level. The CFD analyzes instead of the original detector pulse, a summed pulse shape consisting of the original pulse and a inverted and delayed version of the original pulse (see Fig. 3.19, [78]). The CFD parameter defines the input trigger level of the incoming pulses and the ZC level defines the zero crossing point of the summed pulse shape. One can see in Fig. 3.19 that the amplitude can be triggered in the ZC level range with almost no amplitude jitter and therefore no additional time jitter.

However, the TCSPC electronics requires a minimum pulse width, to be triggered at the input channels. The "Simple-Tau-130-EM-DX" from Becker & Hickel requires for example a minimum pulse width of 400 ps. Increasing the discriminator level sets the trigger level to a reference point on the amplitude slope, which on the other side reduces the pulse width. Therefore, the TCSPC electronics can only measure within certain limitations. Moreover, not all investigations gain from the correction of the measurement by the CFD because it partly fades out the intrinsic characteristic of the detector pulse amplitude.

A typical TTS/jitter measurement of an SNSPD is given in Fig. 3.20 [TGS$^+$12]. For this detector a 5 nm thick SNSPD is used with a nominal line width of 100 nm and an area of $4 \cdot 4\ \mu m^2$. The measurement is done at 6 K. The measured jitter accords to $t_{j,total}$ and therefore, it is the convolution of several system jitters. It does not represent the minimum value that could be reached with SNSPDs.

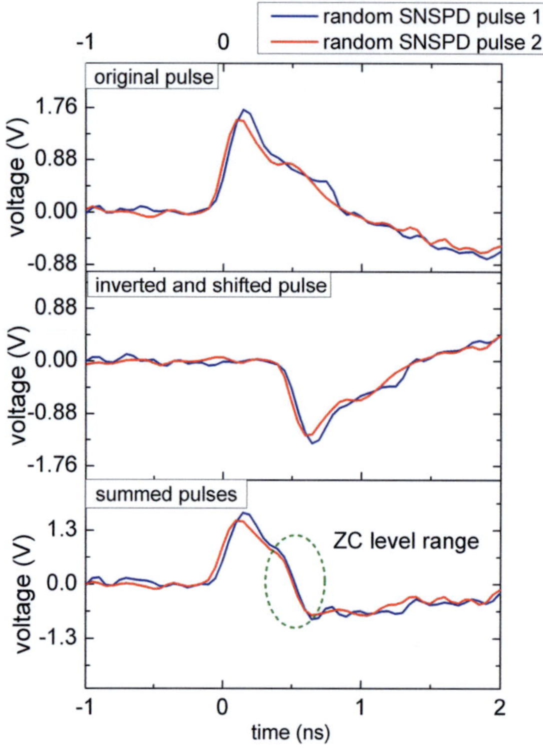

Figure 3.19: Principle of the CFD applied to an SNSPD pulse. In the first graph two random single-shot pulses (red and blue) of an SNSPD and in the second graph the inverted and delayed version of the same pulses are shown. The summed signal of the two pulse versions in given in the third graph. The dashed circle marks the region, where a suitable selection of the ZC level allows measurements without amplitude jitter (red and blue curve almost overlapped).

3.7.2 Improved jitter measurement by a 10 GHz analog front end

Measurements have shown that for SNSPD measurements, oscilloscopes with high bandwidth deliver better t_j values than TCSPC electronics [16], and it has been shown that the bandwidth of the complete readout system influences the quality of a jitter measurement [16]. For use in real applications, a TCSPC electronics is more suitable, since only the pure measurement of τ is often required for data analysis, which makes the measurements less complex. Therefore, an optimized input interface for the TCSPC electronics with higher bandwidth and the possibility of triggering the pulses with a discriminator level comparable to an oscilloscope is helpful. The improved electronics should enable a more accurate triggering of the pulse edge leading to an improved resolution of the jitter $t_{j,\,det}$. In this section, a 10 GHz analog front end is described that enables an amplitude depending selection of detector pulses including an adaption of the pulse shape to the TCSPC electronics [79]. The analog front end has to cover the following requirements:

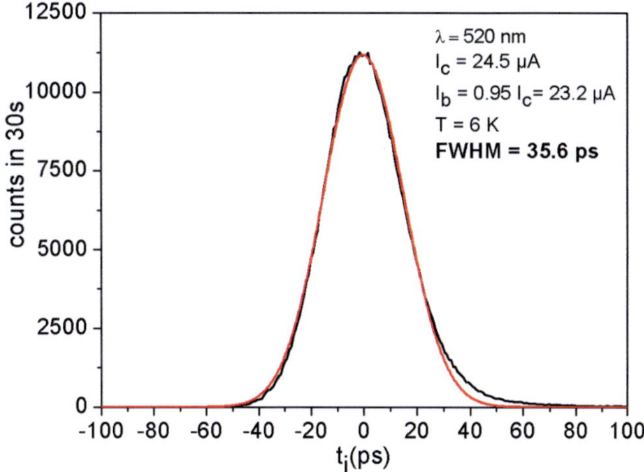

Figure 3.20: Jitter measurement of a 5 nm thick SNSPD at around 6 K operation temperature. The FWHM of 35 ps is extracted from the Gaussian fit (red curve). [TGS$^+$12]

- A frequency input bandwidth of more than 5 GHz to trigger the pulse with the real edge steepness.

- A discriminator level functionality to select a suitable trigger point on the pulse slope.

- A pulse broadening functionality to adapt the short SNSPD response to the lower bandwidth of the TCSPC electronics without increasing $t_{j,\,det}$.

- Mutual channel blocking functionality to select exact start and stop events to trigger on defined pulses.

Design of the analog front end

Based on these requirements, a high-speed analog front end (AFE) is developed. The block diagram of the AFE is shown in Fig. 3.21. The comparator in Fig. 3.21 is the HMC675 from Hittite with a bandwidth of 10 GHz. This comparator has an intrinsic jitter of 2 ps, which is sufficiently low for SNSPD readout. The task of the comparator is to define a trigger level to set a threshold for the incoming pulses on the "+" port. The selected comparator device covers an input range of ±2 V, which is in the range of typical SNSPD pulses after amplification. The comparator switches the output voltage as soon as the input amplitude reaches the reference voltage level which is applied at the "$-$" port of the comparator. Depending on this reference voltage, the pulse can be triggered between zero and the maximum amplitude as long as the maximum input voltage is not exceeded. The output signal of the comparator is differentially connected to the clock input of a 26 GHz toggling flip-flop (HMC679 from Hittite with an intrinsic jitter of 2 ps). The ports of the two components are both "Current Mode Logic" (CML) type. Therefore, no signal conversion is required. The flip-flop toggle input itself is set to 1. That means, if an SNSPD slope exceeds the adjusted trigger level of the comparator the comparator switches, which is interpreted as clock event by the flip-flop. The flip-flop, which output state is always on zero in the stationary state, toggles to 1. The flip-flop has automatically to toggle back to zero after each pulse event. The time length of the flip-flop pulse should match the minimum required pulse width of the TCSPC electronics input. To keep this requirements, the pulse length should be defined flexible without changing any components on the board. Therefore, the non-inverted output signal of the flip-flop is split and feed back as single-ended signal to the reset input of the flip-flop (The reset has actually a differential input, but can also be driven single-ended, if the second input is set to the correct common voltage offset). The pulse length at the output of the flip-flop is defined by the delay of the feed back path. The feed back line is interrupted by two SMA plugs, which allow to connect a separate rf cable in between with individual length. This length provides the dedicated traveling time which delays the reset.

In the TCSPC measurement both TCSPC input channels would gain from adding an AFE in front of the channel ports because the increased bandwidth improves also the triggering of the reference port measurement. Using two modules, the stop channel should only be activated for a certain time after a start event, to limit the maximum reaction time of the stop channel to one pulse laser period. For this reason the AFE has a "latch enable" input, which can be connected to the Q_n port of the output stage. To adapt the output signal to different logic signal types, an offset can be added to the output signal. The signal amplitudes in the current case conform to the CML standard and can be adapted for the "Becker & Hickel" electronics, as well.

In Fig. 3.22 the final device is shown. The assembled board is mounted in a brass holder.

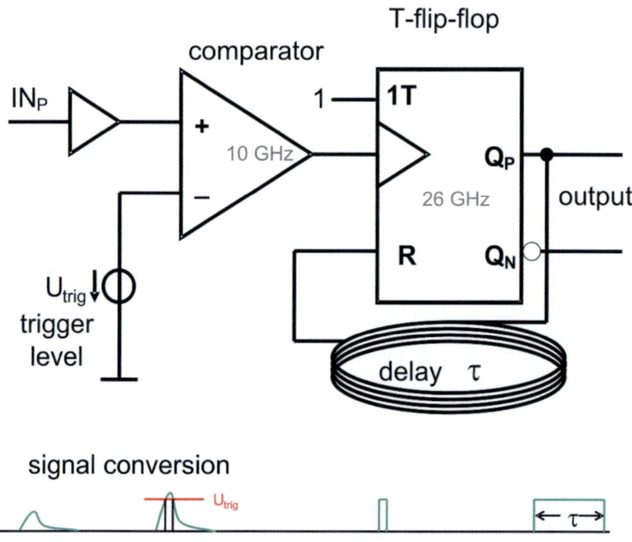

Figure 3.21: Block diagram of the high-speed analog front end and scheme of the pulse conversion

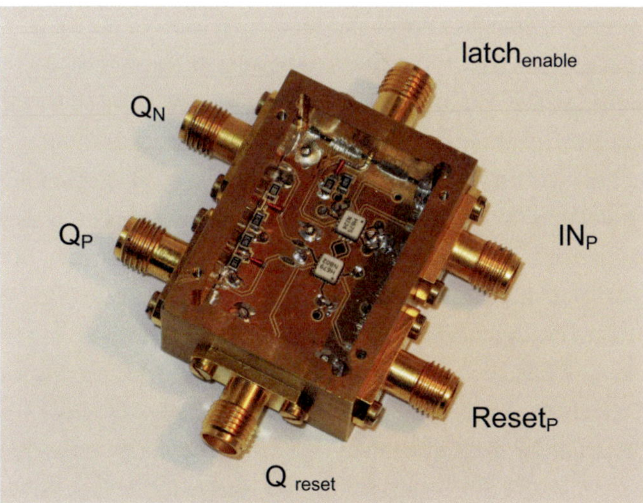

Figure 3.22: Image of the analog frontend module

All dc supply lines are filtered by π-filters. All rf lines are lead trough SMA connectors and therefore, cover a bandwidth of 26 GHz. *IN_P* is the module input and should be connected to the SNSPD amplifier or the reference voltage source. Q_P and Q_N are the non-inverted and inverted output of the AFE and are connected to the TCSPC electronics. Between Q_{reset} and *$Reset_P$* a cable can be mounted, whose signal delay time defines the width of the output pulse. *$Latch_{enable}$* sets the on/off state of the module.

Test of the performance of the AFE and jitter analysis of an SNSPD

The module is tested separately with an HP pulse generator. The system is build like a typical TCSPC setup. The reference pulse is taken from the same output channel of the pulse generator using a power splitter to remove the generator jitter. Both signals are delayed to each other. The reference pulse is directly connected to the TCSPC electronics and the main pulse is connected via the AFE to the TCSPC electronics. The generated pulse is, concerning height and length, similar to an SNSPD pulse after amplification (width: 1 ns and height 0.5 V). Figure 3.23 shows a screen print of the measured jitter, which is measured with a trigger level close to zero. The module reaches a time resolution down to 10 ps, which is close to the resolvability limit of the TCSPC electronics.

One can conclude that the AFE quite well improves the readout of SNSPD pulses. The smallest input pulse, which could be triggered and broadened, had 70 ps, which is much smaller than required from a typical SNSPD pulse width [TGS$^+$12].

A TCSPC measurement of an SNSPD is realized with an AFE mounted in front of the detector input channel of the TCSPC electronics. A 5 nm thick detector is cooled in the dip stick cryostat and biased at $0.95 \cdot I_C(T = 6\ \text{K})$. In the readout line two 3 GHz amplifiers are inserted. The system is illuminated with the 1560 nm fs-laser in the single-photon regime. The reference pulse of the laser is directly connected to the second channel of the "Becker & Hickel" electronics. The AFE trigger level is set to a value, close above the noise floor.

Figs. 3.24 and 3.25 show the extracted t_j values (defined as full width at half-maximum (FWHM)). Once the CFD level is set constant and the ZC level is optimized and once the measurement is done in the contrary way. In a direct measurement, both values have to be optimized to find one operation point which has the lowest t_j value. In the AFE measurement, the variation of the CFD and ZC parameter influence only little the extracted t_j values. The t_j values measured with AFE are close to the best t_j values extracted from measurements without AFE in between. The output pulse of the AFE is quite stable and needs no additional corrections. Moreover, the AFE delivers quite low t_j values without complex parameter optimization, which fastens the preparation time during measurement. The last variable parameter in the TCSPC setup is the trigger level of the AFE, which can be changed by applying an external voltage source at the dc input wire of the second comparator input of the AFE. Figure 3.26 shows the dependence on the jitter to the trigger level of the AFE. The lowest jitter can be kept over a trigger level range of around 60 mV, which is a wide range for practical operation. However, the system first starts to produce reliable results from a trigger level of 320 mV. A reason is that the typical oscillation in the tail of the SNSPD response can be interpreted as new pulse response and adulterate the results. Moreover, the trigger level has to be selected above the noise and distortion floor of the readout system,

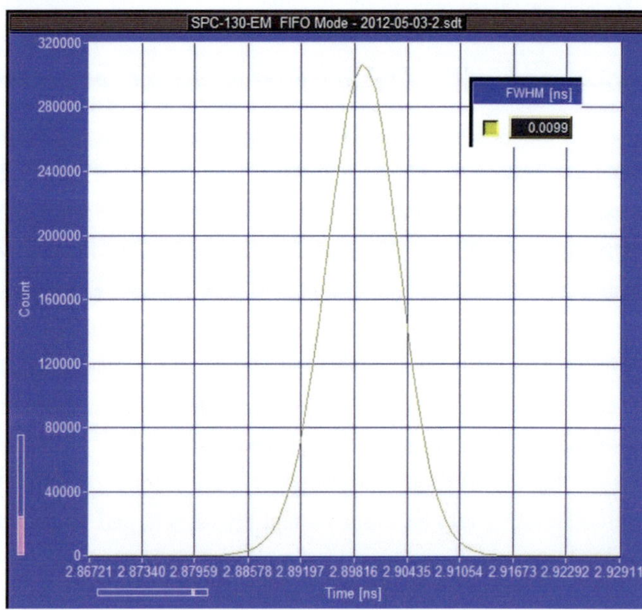

Figure 3.23: Screenshot of the Becker & Hickel controlling software. The measurement shows a 9.9 ps jitter of the AFE board measured in front of the TCSPC electronics. The AFE and the reference port are triggered by an HP pulse generator.

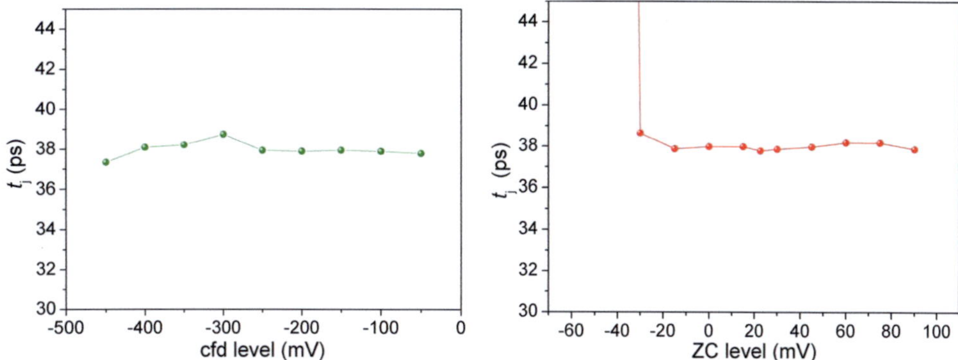

Figure 3.24: Jitter of an SNSPD measured with the AFE depending on CFD level. The ZC level is kept constant.

Figure 3.25: Jitter of an SNSPD measured with the AFE interface depending on ZC level. The CFD level is kept constant.

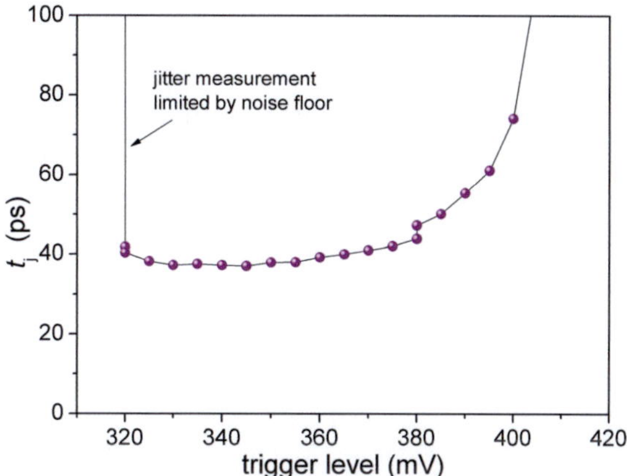

Figure 3.26: Jitter of an SNSPD measured with the AFE interface depending on the trigger level of the AFE. The CFD and ZC level are kept constant.

which now has an enlarged bandwidth of 10 GHz. This disadvantage for the AFE could be reduced by using amplifiers with lower starting bandwidth, lower noise figure and the improvement of the matching between the SNSPD and the amplifier to reduce the general noise and distortion floor of the setup.

One can conclude that the AFE improves the data acquisition of the SNSPD response by increasing the parameter range to enable a minimum jitter and consequently accelerates the optimization process during measurement. Of course, the amplitude jitter of the SNSPD pulse increases the jitter with increasing discriminator level. However, due to the higher input bandwidth of the input stage, the range of discriminator level with minimum jitter is quite enlarged. The pulse broadening functionality and the stable output pulse guarantees that all SNSPD pulse types broader than 70 ps can be measured by the TCSPC electronics. This enhances the use of a TCSPC readout for a wide range of SNSPD geometries.

3.8 Summary

Intense study of SNSPD performance and development of innovative readout concepts require a cryogenic setup which enables fast measurement cycles, low noise operation, high signal integrity and a broad spectral characterization. For this reason, a dip stick cryostat was developed.

The dip stick is optimized to measure SNSPD parameters like detection efficiency over a wavelength range from 400 nm to 2000 nm with an accuracy of 4% and to measure dark count rates. Moreover, the setup enables time jitter measurements of SNSPDs down to 20 ps.

The core of the cryostat is a vacuum stainless steel/copper tube, which contains the detector and its electronics. The detector block is thermally anchored to the liquid helium cooled cryostat tube. The operation temperature can be varied between 4.2K and T_C of the SNSPD. The detector is embedded on a rf carrier board with an integrated self-developed cryogenic biasT allowing a bias current injection close to the detector position at cryogenic temperatures. The full rf system is designed for a 30 MHz to 10 GHz readout bandwidth with ultra-low noise and allows a one channel, two channel readout or transmission measurement, respectively. For biasing of the detector a bias source was developed, which has a current noise $i_{pp} < 400$ pA at 1 kHz and allowing an adjustable current resolution of 0.1 μA. The optical coupling is fiber based by a multimode fiber to realize a broad spectral radiation guiding. The detectors can be irradiated either by a cw halogen light source, whose spectrum is split by a computer-controlled monochromator or by diverse laser sources like a 1560 nm pulsed fs-laser or a 650 nm laser diode. The output signals are recorded by electronics like diverse oscilloscope, the SR620 pulse counter or the "Simple-Tau-130-EM-DX" TCSPC electronics. For TCSPC measurements, a special 10 GHz analog front end (AFE) was developed which enables the TCSPC electronics to operate with a higher input bandwidth than by the pure TCSPC electronics. It works itself with a time resolution of 10 ps and broadens the range of suitable operation parameters of the TCSPC electronics leading to minimum jitter values and enables pulse broadening of the detector response for adaption of commercial TCSPC electronics of pulses with widths down to 70 ps.

4 Improved performance of SNSPDs

A high performance of a single-photon detector system is both determined by the quality of the readout setup in combination with the readout concepts and the understanding of the intrinsic processes in the SNSPD itself. Knowing the exact physical behavior of the detectors allows to optimize the technological fabrication processes and in the same way an improvement of the detector quality. The focused parameters are the detection efficiency and the dark count rate of SNSPDs. The detection efficiency is analyzed in detail concerning several dependencies like film thickness, bias current, wavelength, and is compared to suitable physical models which explain the detection in the infrared range by a vortex contribution. In connection with a pulse response study the plausibility of the models is shown and the similarity between the evolution of dark counts and the mechanism of infrared detection is discussed. The reduction of the dark counts is a further focus of this chapter. Three types of modifications are introduced, which allow a reduction of the dark count rate. These are an improved cooling, optimized stoichiometry and operation temperature variation.

4.1 Detection limits, detector fabrication and analysis of basic detector dependencies.

Until now, not all intrinsic processes in SNSPDs are well understood. The hot-spot detection model [19] itself can not completely describe the behavior of an SNSPD. That means, new or extended physical models are required in order to be tested as suitable explanation for the intrinsic processes. In general, physical investigation can be done on many parameters, like variation in fabrication conditions, superconducting material, substrate selection and geometry variation. The variation of the detector geometry is comparatively easy to realize during the fabrication process. Changes in the geometry have a high impact on superconducting parameters and also electrical characteristics, and can help to demonstrate evaluable dependencies.

In this subchapter the thickness dependence of SNSPDs is investigated. The analysis of several parameters like bias current, wavelength and the investigation of the pulse shapes in dependence on the film thickness enables to improve the understanding of the physical processes in SNSPDs. The focus in this subchapter is set to the intrinsic detection efficiency of an SNSPD, as well, which enables a better view to the intrinsic dynamics in a thin and narrow meander revised from influences like absorption. The section is an extended version of the discussion of the experiments presented in [HRI$^+$10b].

4.1.1 Energy threshold of the hot-spot formation

Before starting with the experiments, an introduction to the hot-spot limitations is required, which is defined as the energy threshold of the hot-spot formation. To understand the energy threshold of the hot-spot formation, we first have to consider the superconducting energy gap more in detail. The superconducting energy gap $E(0)$ describes the energy required for Cooper pair breaking. This energy gap has a temperature dependence and can be described according to the BCS theory for $T \ll T_C$ [18]. At $T = 0$, one defines

$$E(0) = 2 \cdot \Delta(0) = \frac{\pi}{e^\gamma} \cdot k_B T_C = 3.528 \cdot k_B T_C \tag{4.1}$$

Here, $\Delta(0)$ is the energy gap for one of the electrons in the Cooper pairs, $\gamma = 0.577$ is the Sommerfeld constant, $e^\gamma = 1.781$, k_B is the Boltzmann factor and T_C the critical temperature of the superconductor. The energy gap of a superconductor typically lies in the order of several meV. The factor 3.528 can only be treated as an approximation. For example high-T_C superconductors already aberrate from this value [80]. In case of NbN, which is used in this thesis, the definition is valid for bulk material. For thin NbN films, a factor of 2.07 has been shown as suitable parameter [81]. Further investigations concerning the superconducting gap of thin NbN films are analyzed in [HDH$^+$12].
To cover the complete temperature range T between 0 K and T_C one can take the following approximation [47].

$$E(T) = 2 \cdot \Delta(T) = 2 \cdot \Delta(0) \left[1 - \left(\frac{T}{T_C} \right)^2 \right]^{\frac{1}{2}} \left[1 + \left(\frac{T}{T_C} \right)^2 \right]^{\frac{3}{10}} \tag{4.2}$$

In the second chapter the principle of the SNSPD detection mechanism, called the hot-spot model, is derived from the two-temperature model and extended by a electro-thermal model. The hot-spot model describes the detection mechanism, which is relevant for the visible and near UV spectral ranges [10]. The photon, absorbed in the superconducting nanowire, produces a non-equilibrium hot spot where the concentration of Cooper pairs is reduced. The remaining Cooper pairs have to accelerate in order to carry the externally applied bias current. Once the velocity of Cooper pairs reaches the critical value, the superconductivity over the entire cross-section of the wire breaks down and the normal conducting, resistive domain appears and forms a normalconducting belt [82], [18], [19]. Due to this normalconducting barrier, the bias current is redistributed from the meander line to the rf readout line with a certain time constant. Depending on the dynamics of the current the declining part of the bias current in the normalconducting belt starts to dissipate Joule power. Depending on the local strength of cooling, this may lead to a growing or shrinking of the normalconducting belt [20].
Hence, the detection process can be temporally divided into two consecutive stages. The

first one is the absorption process which leads to a local appearance of a hot spot in the superconductive meander line and the second is the evolution of the normalconducting belt. For photons in the UV to near-infrared range the amplitude and the duration of the voltage transient are both defined at the second stage of the response. Eventually, the dynamic of the normalconducting belt resistance is defined by the physical processes given by the thermal relaxation time τ_{es}, the electrical time constant τ_{el} of the electrical readout circuit and the impedance of the detector [27].

In general, the appearance of the resistive belt needs a certain activation energy because a certain number of Cooper pairs must be broken to suppress the superconductivity in the meander. However, depending on the absorbed photon energy, a hot spot can even exist in a meander without leading to a normalconducting belt, since the hot-spot itself, which produce a partly resistive area on the meander line does not prevent a current flow in the detector path as long as no complete normalconducting belt exists. Therefore, the absorbed energy needs to be sufficient in order to come from the first stage of detection, the hot spot, to the second stage, the normalconducting belt. The activation energy should correlate with a dedicated minimum absorbed photon energy. That means, if the minimum required activation energy is known, there must be a dedicated limit in the spectral response of the SNSPD at a certain wavelength λ_0, the so-called cutoff wavelength. The exact boundary depends, amongst others, on the cross-section of the wire and is therefore geometry dependent. To get a condition for evolution of the normalconducting belt concerning the required photon energy, one has to set the number of required broken cooper pairs to get a normalconducting belt equal to the number of available broken Cooper pairs of a photon absorption [21].

During operation, a detector is usually biased close to I_C. In the superconducting state the difference ΔI between the operation bias I_B and the critical current I_C defines a certain portion of energy $\delta\Delta$. The energy $\delta\Delta$ is required to set the meander to the normalconducting state. As long as the meander is still superconducting, a certain number of Cooper pairs still exists, which act as the charge carriers of the current transport. This $\delta\Delta$ can be defined in dependence on the Cooper pair density n_s:

$$\delta n_s / n_s = 1 - I_B / I_C \tag{4.3}$$

Here, n_S is the equilibrium density of Cooper pairs (based on the operation temperature and the bias current) around the photon absorption area and δn_S is the change of current imposed by the absorbed photon. The density of Cooper pairs n_s can be derived from the electron density of states N_0 and the superconducting gap Δ at operation conditions. Assuming that the hot spot has as a minimum length of the coherence length ξ, the minimum initial hot-spot volume is defined by $V = \xi w d$ (w is the meander width and d the film thickness). The number of required broken Cooper pairs (called quasiparticles) can therefore be approximated from Eq. 4.3:

$$\delta N_{qp,c} = n_s \xi w d (1 - I_B/I_C) \tag{4.4}$$

The absorbed photon energy ε_{ph} is diffusing from the initial point with a certain diffusivity constant D for a thermalization time τ_{th} which can be interpreted equal to τ_{e-ph}. The dedicated volume is defined by $V_{therm} = \sqrt{\pi D \tau_{th}}$. The complete photon energy is not available over the complete hot-spot evolution time τ_{th} because a certain part already recombines or dissipates. Therefore, the available number of quasi-particles is reduced to $M(\tau_{th}) = \varsigma \varepsilon_{ph}/\Delta$. The parameter ς denotes this as effectiveness of the energy transfer from the absorbed photon to the electrons in the superconducting NbN film. The maximum number of available quasi-particles has to be integrated to the dedicated maximum volume V_{therm} that the hot spot will reach at τ_{th}.

$$\delta N_{qp} = M(\tau_{th}) \cdot \frac{\xi}{\sqrt{\pi D \tau_{th}}} \tag{4.5}$$

Setting Eq. 4.3 and Eq. 4.5 equal and one can solve this equation to get a definition for the photon energy ε_{ph}, which is required as activation energy for the appearance of the resistive belt [21]:

$$\varepsilon_{ph} = \frac{hc}{\lambda_0} = \frac{3\sqrt{\pi}\Delta^2 w}{4\varsigma e^2 R_S} \sqrt{\tau_{th} \cdot D} \left(1 - \frac{I_B}{I_C^d} \right) \tag{4.6}$$

R_S is the square resistance and replaces the density of state n_s, which is difficult to measure. I_C^d is the depairing critical current.

For smaller photon energies or, equivalently, for wavelengths larger than the cutoff wavelength λ_0, a steep drop of the detection efficiency. This prediction will be analyzed with the subsequent experiment in this chapter.

4.1.2 Detector fabrication and characterization

In this section the detector fabrication process is described. The fabrication steps are identical for all characterized detectors in this thesis. Therefore, the fabrication is described only at this position and referenced in the other chapters. If one experiments needs special fabrication steps, this will be explained separately in each dedicated section. Moreover, the fabrication process is described in more detail in [HDH$^+$12]. In general, the detector fabrication required for all measurements in this thesis is done both by Konstantin Ilin and Dagmar Henrich. The enhancements of masks and detector designs are all realized by the author. Some deviating tasks are referenced individually.

NbN thin film deposition

The material of the superconducting thin film is NbN. NbN is a classical type-II superconductor. The typical bulk critical temperature is $T_C = 17$ K. The type-II characteristic is given by the Ginzburg-Landau parameter $\kappa = \lambda_L/\xi \gg 1$. λ_L is the magnetic penetration depth and ξ is the superconducting coherence length. The typical substrates for the SNSPD fabrication at IMS is sapphire. The advantages are the good crystalline lattice matching of NbN and sapphire which allows higher T_C values in thin films. Indeed, other substrates are possible like MgO [83] or SiO_2 [84], as well.

The NbN films at IMS are deposited by dc-reactive magnetron sputtering of a pure niobium target in an Ar and N atmosphere onto R-plane sapphire substrates. Since the cryogenic setup illuminates the detectors from the top side, one-side polished substrates suffice for the experiments. The Nb target is pre-cleaned by sputtering in a pure Ar atmosphere before deposition. The substrate is kept at a temperature of 750 °C during deposition. The Argon + N_2 mixture is ionized by a magnetron. On the surface of the target the reaction of Nb and N leads to NbN molecules, which are sputtered from the target by accelerated ions. The NbN is deposited on the heated sapphire substrate. By changing the discharge current the ratio between Nb and N can be defined [HDH$^+$12].

The film thickness for all experiments in this thesis is selected to be between 4 and 12 nm. The thicknesses, which are presented in Table 4.1, are calculated from the sputtering time and the known deposition rate (between 0.06 and 0.17 nm/s depending on the discharge current). The thickness is additionally confirmed by ellipsometric measurements.

Single pixel detector design with coplanar readout

The core of the SNSPD is a narrow meander structure. The meander patterning is made with positive resist by electron-beam photo lithography. The readout line on the detector chip is designed as coplanar waveguide and fabricated by standard photo lithography. The structures are etched by Ar^+ ion-milling and CF_4 and SF_6 plasma, respectively [HDH$^+$12].

The meander lines have a width in the range 90 ± 15 nm and a nominal filling factor of 50 % (the ratio of the width of the NbN wires to the sum of wire width and gap between the wires in a meander design). The meander pattern covers an area of 4x4 μm^2, which defines the active area of the detector. The actual widths and lengths of the meander lines are measured by means of scanning electron microscopy (SEM) (see Fig. 4.1) with a resolution better than 10 nm. For each meander the line width is measured at a few locations along the line. The widths, which are listed in Table 4.1, are the mean values obtained from these measurements. Already at this step of device fabrication only the meanders are accounted which have no visible defects (constrictions, breaks, shortcuts, etc.) of the meander wires.

The experiments in this section are done with an older design of the coplanar waveguide (see Fig. 4.2 a)). The older design originates from the fabrication of hot-electron bolometers,

Figure 4.1: SEM image of the geometry of a detector structure after etching. The bright lines are the etched gaps between the meander stripes.[HRI+12]

which are illuminated from the back side. During this thesis, the design is optimized for a 50 Ω SNSPD readout. The coplanar parameters are calculated for a substrate with a GND plane on the bottom. Moreover, the contact between the end of the meander and the surrounding GND plane is minimized to a minimum loop way. The dimensions for the coplanar line (see Fig. 3.5) were calculated by Microwave Office to the following parameters according to Fig. 3.5: loss-less film resistivity, width $w = 500$ µm, gap $g = 550$ µm, $h = 300$ nm thick substrate with $\varepsilon_r = 10$, tan $\delta = 0.0025$ (sapphire). The final coplanar design (see Fig. 4.2 b)) is patterned on the NbN film by a HR mask with contact photo lithography.

Fig. 4.3 shows a SEM image of the final meander line with its contact leads. The leads broaden the line width to the line width of the coplanar waveguide, which is outside of the image view.

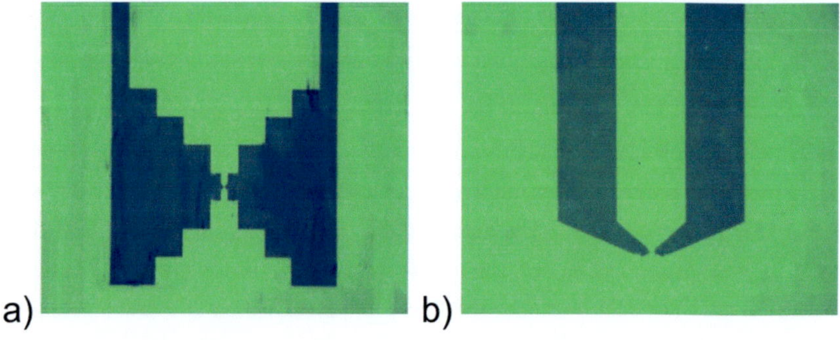

Figure 4.2: a) Old coplanar readout design. The detector is in the middle of the two contact triangles. b) New coplanar readout design.

Table 4.1: Overview of the characteristic parameters of the analyzed detectors. d is the thickness, w the meander width, I_C is the measured critical current at 4.2 K, R_S is the square resistance and λ_0 the cutoff frequency. [HRI$^+$12]

detector	d(nm)	w(nm)	T_C(K)	I_C(µA)	R_S(Ω)	λ_0(µm)
1	4	89	11.7	18	700	0.76
2	4.3	80	11.45	18	501	0.57
3	5	80	12.2	31	432	0.40
4	5	97	12	47	391	0.44
5	5.7	76	10.7	18	446	0.63
6	7	93	12.8	70	274	0.27
7	8	94	14.3	106	191	0.13
8	8	74	12.7	21.5	402	0.37
9	9	92	14.3	122	173	0.13
10	10	103	14.9	186	131	0.08
11	12	91	13	160	109	0.07
12	12	99	14.9	220	104	0.09

Measurement of critical temperature and critical current.

Within a short time after the deposition, the critical temperature T_C of each film is measured in a separate cooling cycle. T_C is the temperature, which is measured, when the film resistance is decreased to 1% of its value above the superconducting transition ($R_{20\,K}$).

T_C shows in Fig. 4.4 a significant reduction with the decrease of the film thickness. This phenomenon, which is typical for superconducting films with thicknesses of a few coherence lengths, [85],[86] can be explained by the proximity effect [87], [88], [89] and the decrease in the electronic density of states with increasing disorder [39].

After patterning, the superconducting resistive transition T_C and the critical current I_C at 4.2 K are again measured for all meanders in each series (nine meanders with the same

Figure 4.3: Etched meander line connected to the waveguide taper.

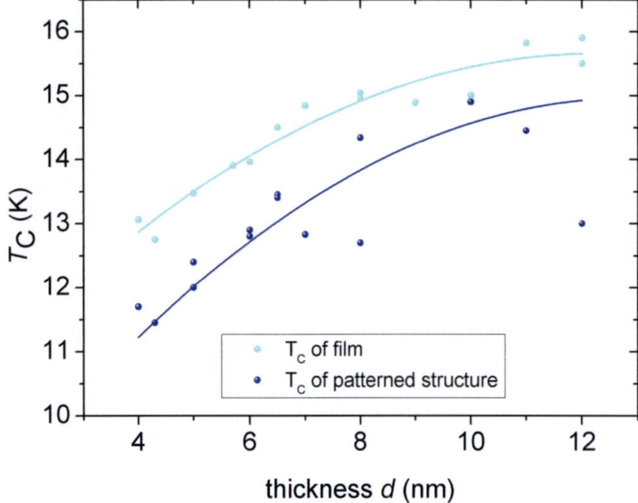

Figure 4.4: Dependence of T_C on NbN film thickness. The films are measured before and after the meander patterning. The T_C shows the typical decrease with decreasing thickness, which is explained by the proximity effect. The patterned films have T_C of about 1 K less than the non-patterned films.

nominal thickness on a $10x10$ mm^2 chip) in a separate cooling cycle. The critical temperatures T_C of the meanders are 1 to 2 K lower than T_C of the corresponding non-patterned films but show a similar dependence on the film thickness (see Fig. 4.4)[90].

The critical density j_c over thickness is depicted in Fig. 4.5. It has the same decay with decreasing thickness, as the critical temperature. Nevertheless the critical current I_C of all

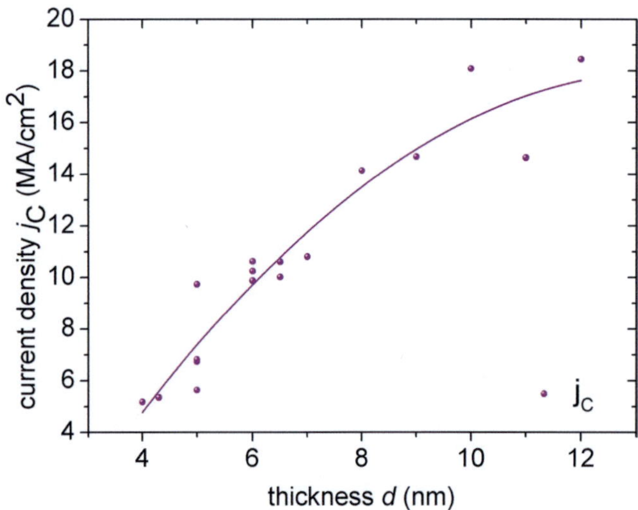

Figure 4.5: Dependence of the critical current density j_C on film thicknesses d.

these detectors is large enough at the cryostat temperature (6 K) providing sufficient pulse amplitude for counting single-photon detection events.

For suitable selection of a detector from one NbN film, the resistance per square can support the evaluation of the detector quality. The resistance per square is defined for each meander as $R_S = R_n \cdot (w/l)$, where R_n is the total resistance of the meander above the transition (R_{20K}) and w and l are the width and the total length of the meander lines both extracted from the SEM image. Although the spread of the R_S values within each detector series is less than 20 %, for optical measurements the optimal meanders are selected which have the smallest ratio of R_S to the critical current density j_c.

4.1.3 Free-space optic bath cryostat

The scheme of the experimental setup is shown in Fig. 4.6. In contrast to all other experiments in this thesis, this experiment is realized in a bath cryostat of DLR Berlin with some different electronics. Therefore, the setup is explained in detail.

The broad spectral light comes from a halogen lamp and is passed trough a manually operated monochromator, which enables measurements in the spectral range from 400 to 2000 nm wavelength. The light at the output of the monochromator is polarized parallel to the detector meanders. The light beam after the monochromator is expanded to a diameter of several centimeters that allows homogenous illumination of a few millimeters large objects in the center of the beam. For optical measurements, the meander is positioned at the center of the beam and illuminated at normal incidence from the top side of the detector. To

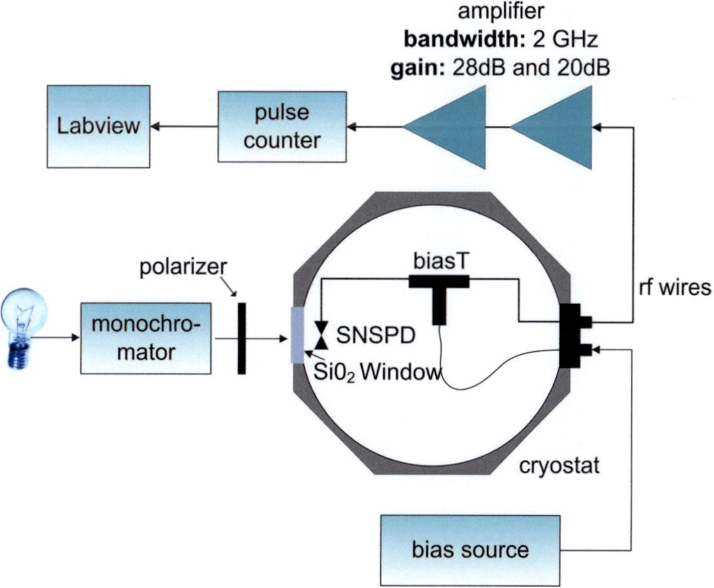

Figure 4.6: Experimental setup of a free–space coupled bath cryostat.[HRI$^+$12]

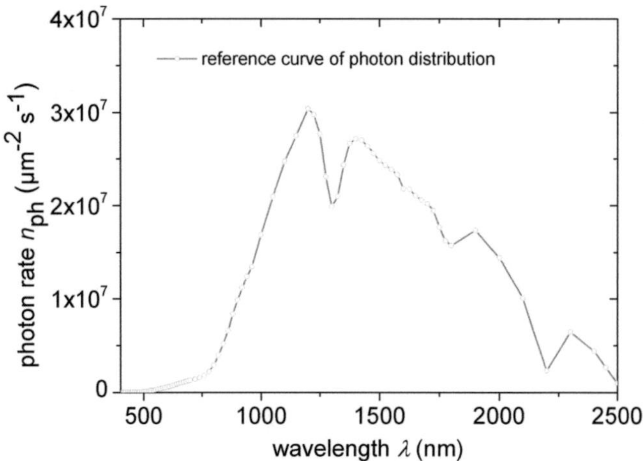

Figure 4.7: Dependence of photon rate per μm^2 of the halogen light source on wavelength. The data is extracted from the calibration process, which is accomplished before the optical measurements.

attain the absolute numbers of photons for calculation of DE, the light source is calibrated before measuring the photon count rates. The light intensity is measured with three certified photo diodes which cover with a certain overlap the full wavelength range:

- Si diode: 400 nm to 1000 nm

- Ge diode: 800 nm to 1500 nm

- PlSe diode: 1000 nm to 3500 nm

The overlap is necessary because the sensitivity of the edge of the spectral bandwidth is always quite poor and produces errors in the calibration. Moreover, some of the diodes only deliver voltage values correlated to the incident photon flux, but no absolute power values. Therefore, the measured values need an absolute power reference. During calibration the diodes are placed at the exact meander position after an iris diaphragm with a diameter of less than 1 mm. At this position the light distribution of the beam on the diaphragm can be assumed to be homogenous. The losses of the quartz window are also measured and accounted with 8.8 %. The experiment is executed only with polarized light parallel to the meander lines. The polarization of the incident light is controlled by a polarizer which is installed between the monochromator and the cryostat window. The diodes deliver a certain optical power for each wavelength. This power is converted into a photon number rate for a defined illumination area. Fig. 4.7 shows the photon distribution over wavelength. The light source has the maximum in the near infrared range. The wavelength range between 400 nm and 2000 nm is quite well covered. Measurements close to 400 nm can have some uncertainty because of the low photon number. The monochromator allows to change the

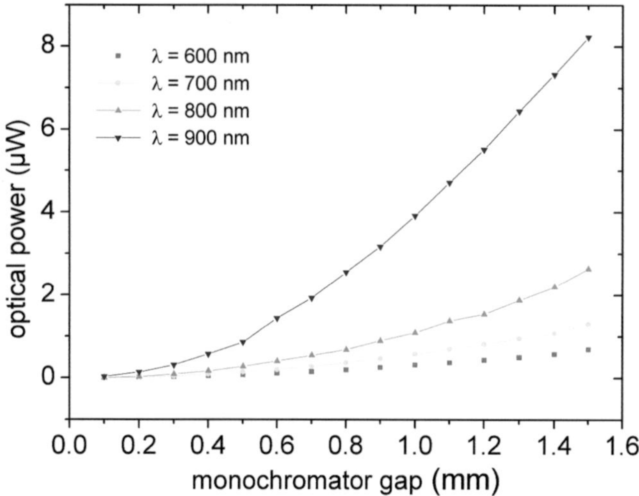

Figure 4.8: Dependence of the optical power on gap width of the monochromator output and on wavelength.

optical output power by varying the output gap of the monochromator. This is helpful during the measurement to get always sufficient counts for significant statistics. Since the light intensity does change exponentially with increasing gap, the optical power is measured for each gap size to enable later correction in the experiment (see Fig. 4.8).

For the measurements of the detection efficiency (DE) the detectors are cooled down to approximately 6 K in the vacuum chamber of a He^4 bath cryostat with optical access through a quartz window. The detectors are mounted on a copper holder which is thermally anchored to the 4.2 K plate of the cryostat. The thermal load of the detector is minimized by anchoring all electrical cables to the cold plate. The SNSPD is biased by a low noise voltage source with a cryogenic biasT. The voltage transients, which are generated by the detector in response to single photons, are guided out of the cryostat by a coaxial cable. The signal is amplified by two rf amplifiers with a total gain of 48 dB and a bandwidth 10 MHz to 2 GHz and is evaluated by the SR400 pulse counter (bandwidth 300 MHz, time resolution 5 ns), which is controlled by Labview. After amplification, the transients have a duration of 6 ns that limit the maximum count rate to $1.6 \cdot 10^8$ s^{-1}. The time resolution of the counter is in the same range. Consequently, it is ensured that the number of counts acquired by the counter always correspond to the number of transients released by the detector. The accuracy of the measurements is most strongly affected by the variation in time of the photon count rate. To improve the accuracy, the counts are measured for a few seconds (the so-called gate time) at each set of experimental parameters.

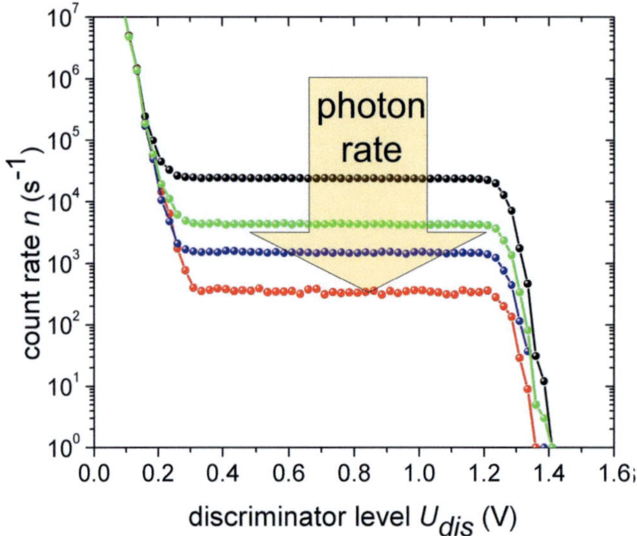

Figure 4.9: Count rate depending on the discriminator level for different arbitrary photon rates.

4.1.4 Study of the detector response and analysis of the detection efficiency

The detector performance is studied for several parameters to improve the understanding of the intrinsic detector processes. In the following, the discriminator level also called trigger level and the amplitude distribution of a pulse response and moreover, the influence of the thickness on amplitude and the dependence of the detection efficiency on bias current is analyzed.

4.1.4.1 Study of the SNSPD response concerning spectral bandwidth and thickness

Since the intensity of light is measured in an SNSPD by the number of pulses and not by the shape of the pulse, the response amplitude is often neglected as long as the response is high enough. However, the pulse response of an SNSPD contains already some information about the intrinsic processes in the detector.

The pulse shape is mainly defined by the second stage of the detection model as described in section 2.1.1. A statistical analysis of the amplitudes can be done by measuring the count rates of a detector with different discriminator levels of the counter electronics, which means a variation of the trigger level. One should be aware of the reduced bandwidth of the pulse counter of 300 MHz, which behaves as a low pass filter to the pulse amplitudes contrary to a higher resolved real-time oscilloscope screen-shot. Indeed, the typical attributes of the amplitudes can be seen and analyzed.

Figure 4.9 shows the measurement of the count rate on the discriminator level of a 4 nm thick SNSPD at $I_B = 0.95 \cdot I_C$. The detector is lighted by different arbitrary photon rates. Until around 500 mV, the count rate is dominated by the noise floor of the readout system for all photon rates. Each photon rate leads to a broad plateau. For lower photon rate this plateau of constant counts shrinks to a lower level. For low count rates it is quite important to select a discriminator level not too close to the noise floor. Increasing the discriminator level one can see that statistically, the counts do not drop with a sharp bend at a certain discriminator value. The transition is a smooth process over a wider discriminator level range. This shows that the amplitudes of the SNSPDs are not identical, but have a certain amplitude distribution. According to the hot-spot model the second stage of the detection, which defines the amplitude shape, can produce different amplitude depending on the current distribution in the meander and the material composition at the photon absorption side. So from detector side, a broader distribution of the amplitude would mean a higher inhomogeneity of the meander.

The discriminator level measurement gives a statistical overview about the amplitude distribution and mean amplitude values. However, the 300 MHz bandwidth of the SR400 counter limits exact conclusions. Therefore, a separate amplitude analysis is required including an amplitude jitter measurement with higher frequency bandwidth of the readout.

The amplitude analysis is done with a 4 nm thick standard sample of the series of section 4.2.4. For the measurement, an amplifier with an effective bandwidth of 1.7 GHz and a high-speed oscilloscope is used to reproduce the original pulse shape of the detector as good as possible without strong amplitude shaping by the readout electronics. The detector is biased with a bias current $I_B = 14\ \mu A$. The pulse shape can be seen in Fig. 4.10. The quite low bias current I_B leads to a low SNR. The noise floor already strongly influences the pulse shape. The amplitude distribution of the measured pulses is depicted in Fig. 4.11 for three different wavelengths in the optical and infrared range. The amplitude distribution is also quite noisy, due to the low SNR. Therefore, for all three measurements, a Gaussian fit is put on the data. The mean pulse amplitudes A_0 described by the maximum of the Gaussain peaks increase with increasing wavelength. The *FWHM* value represents a measure of the amplitude jitter of the detector. The *FWHM* decreases slightly with the wavelength. The high bandwidth measurements are in accordance with the measurements of reduced bandwidth in [52]. One can conclude that the pulse amplitude shifts with the photon energy. Assuming that the photon energy only influences the first part of the detection mechanism and not the growing and shrinking of the resistive belt, one should expect from the hot-spot mechanism the amplitude to be independent on the wavelength. Although the differences are quite marginal, one should doubt that the hot-spot model is a suitable model to explain the pulse evolution in the infrared regime, as well.

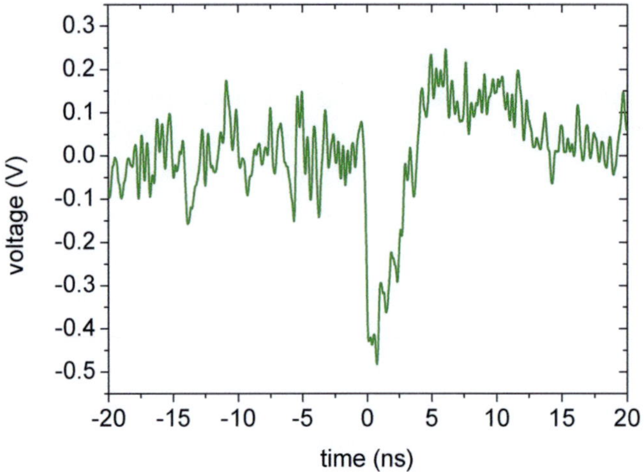

Figure 4.10: SNSPD pulse of a 4 nm thick NbN detector at $I_B = 0.9 \cdot I_C = 14$ µA.

In the next step, the pulse responses are analyzed with a 2 GHz real-time oscilloscope for different film thicknesses. There are mainly two varying parameters that influence the pulse shape, if the thickness changes. The square resistance R_S decreases with increasing thickness d and the critical current I_C increases and consequently I_B, as well. The square resistance R_S and the size of the normalconducting belt determine the absolute size of the resistive part in the meander. Depending on the relation between the resistance in the meander and the 50 Ω of the readout, the applied bias current is split between these two paths (see Fig 5.6).

In Fig. 4.12 the pulse shapes for different thicknesses are plotted. All detectors are biased at $0.95 \cdot I_C$ and illuminated by incoherent light with $\lambda = 500$ nm. In Table 4.1 one can see that I_C increases with increasing d, which would lead to an increase of the pulse amplitudes. R_S and the normalconducting belt resistance, in contrary, decrease with increasing d. Since the change in current should linearly influence the amplitude and the change in R_S only changes the distribution of current in relation to the 50 Ω of the readout impedance, the increase of I_C and consequently I_B should dominate the amplitude change. Fig. 4.12 supports this expectation. The pulse amplitudes increase with thickness, but not in the same amount than one would expect from the increase of I_C and I_B, respectively. The reduction of R_S limits the pulse shape growing.

The analysis of the pulse length is a crucial feature to evaluate intrinsic processes, as well. However, the complete readout path has a bandwidth between 10 MHz and 2 GHz. This is actually not enough to depict the complete real pulse shape. Especially, the length is strongly influenced by the lower bandwidth limit of the readout path. Therefore, an analysis of the decay times is mostly dominated by the readout electronics and does not represent the real pulse shape.

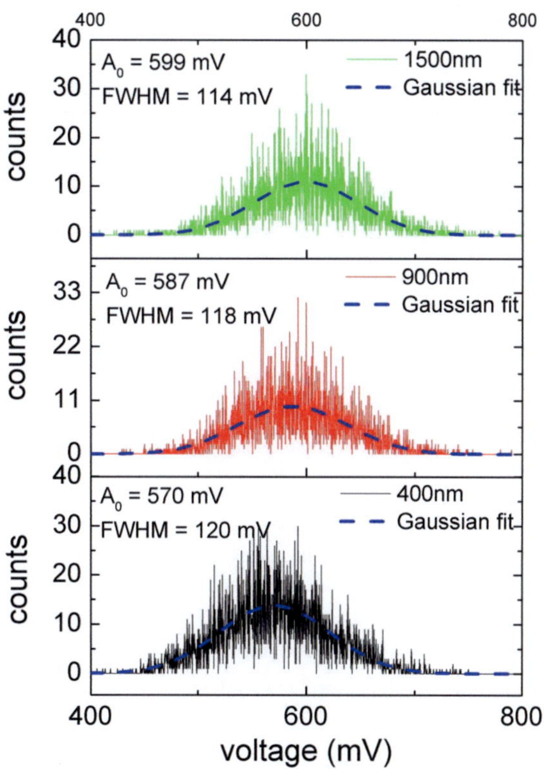

Figure 4.11: Amplitude distribution of SNSPD pulses of a 4 nm thick film at $I_B = 0.9 \cdot I_C$ at different wavelengths recorded by the 32 GHz oscilloscope.

4.1.4.2 Dependence of count rate and detection efficiency on bias current, wavelength and thickness

The following analysis has been published in [HRI$^+$10b].

We want to analyze the count rate and the detection efficiency, respectively, concerning different parameters to deepen the discussion about internal processes in the detector.

In Figure 4.13 the total count rate at different bias currents for the meander 4 from Table 4.1 is given. The bias current I_B on the x-axis is normalized to the measured critical current I_C, the dedicated operation temperature is ≈ 6 K. The photon flux on the detector is kept constant at $\lambda = 500$ nm.

The count rate rises up, the more I_B approaches the critical current I_C. The strong rise of the counts in the range close to I_C is caused by dark counts (see section 2.2.2), which dominate for $I_B > 0.97 \cdot I_C$. This behavior is typical and could be recognized for all investigated detectors.

Since the dark counts influence the accuracy of the DE measurement in the current range close to I_C, the dark counts are accounted by subtracting the mean value of dark counts from

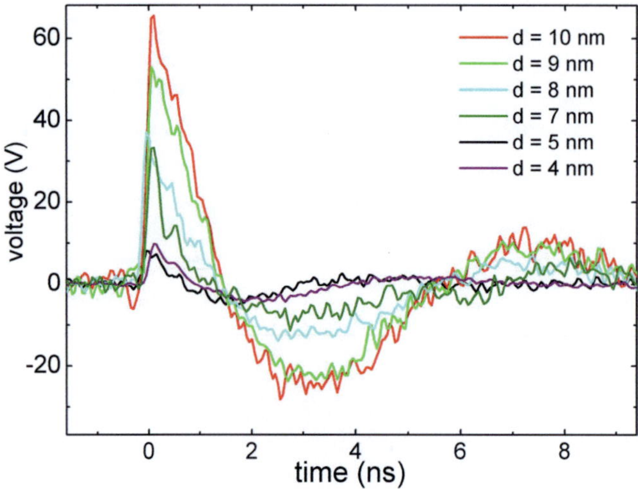

Figure 4.12: Pulse shape for different thicknesses at incident light of $\lambda = 500$ nm. The detectors are biased at $0.95 \cdot I_C$ and pulses are measured by a 2 GHz real-time oscilloscope. The ringing of the pulse shapes mainly comes from the noise floor of the readout system.

the total number of counts. The dark count rate is defined by a separate measurement for each detector in the same setup with complete blocked cryostat window with light-absorbing material. With this method both, the intrinsic dark counts and the black body radiation based counts can be accounted for data correction. All measurements have to be done in the single-photon regime, for which reason the photon count rate is checked to be proportional to the incoming power of radiation for all detectors covering the range of all applied currents and

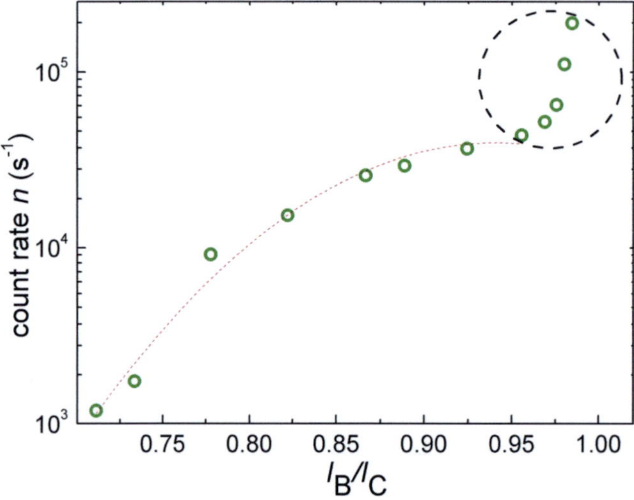

Figure 4.13: Dependence of the counts of an SNSPD on the bias current I_B at $T = 6$ K. The dashed circle marks the range, where dark counts dominate the count rate. [HRI⁺10b]

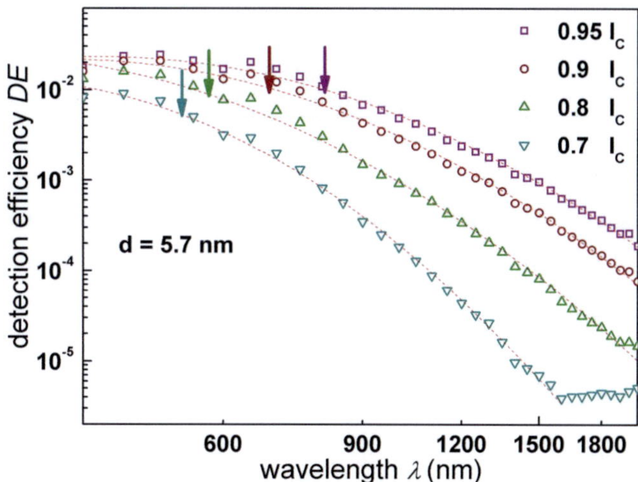

Figure 4.14: Dependence of the detection efficiency on the bias and wavelength at $T = 6$ K. The arrows mark the cutoff of the detection efficiency without correction of the *ABS*. [HRI$^+$10b]

wavelengths. For the evaluation of the *DE* with the absorptance data from section 2.2.1 only the polarization parallel to the wires is considered, since this polarization provides a much larger *ABS* of the meander at infrared wavelengths. The photon flux at the meander position is calculated for each wavelength by using the calibration data of section 4.1.3. Moreover, the area of the meander is required, which is taken from a SEM image, to define the exact number of photons crossing the active meander area during the gating time and the *DE* is obtained by dividing the measured count rates by the so calculated photon number. Taking the approximated values of the *ABS* and the calculated values of the *DE* the intrinsic detection efficiency *IDE* can be defined.

In Fig. 4.14 *DE* of the 5.7 nm thick meander (sample 5) at several I_B is depicted. The *DE* varies little at wavelengths less than 700 nm and decreases monotonically at larger wavelengths for $I_B = 0.9 \cdot I_C$. This short-wavelength plateau is depending on the applied bias current and shrinks with decreasing I_B until at $I_B = 0.7 \cdot I_C$ it disappears completely.

Figure 4.15 depicts the normalized detection efficiencies for meanders with varying thicknesses d. All detection efficiencies are acquired at $I_B = 0.95 \cdot I_C$ and are normalized to the detection efficiency at the wavelength 400 nm. The two detectors with thicknesses 4 and 5 nm have an increase of the *DE* depending on wavelength in the visible range. This phenomenon is due to the increase of the *ABS* with wavelength (see section 2.2.1), which increases significantly for all thicknesses at wavelengths between 400 and 1000 nm (see Fig. 2.5). A general behavior is that above 800 nm the *DE* decreases monotonically for all detectors. The steepness of the monotone decline slightly varies with d.

The measured detection efficiencies have obvious cutoffs as predicted in section 4.1.1, which start in the wavelength range estimated with Eq. 4.6. But the subsequent behavior differs from the expectations. Instead of a steep dropping, the *DE* monotonically decreases (Figs. 4.14 and 4.15) with a certain steepness with increasing wavelength. Since the wavelength dependence of the absorptance (*ABS*) has a similar course for all meanders (Fig. 2.5), one can assume that the positions of the cutoffs behave, relatively considered, similar for different thicknesses. The arrows in Figs. 4.14 and 4.15 mark the position of the cutoff wavelengths (defined as 3-dB cutoff), which are in case of the thickness dependence obtained by correcting the data by the wavelength dependence of the *ABS*.

A comparison of the experimental data of the *DE* with detection models requires the correct ratio of the I_B and the analytical depairing critical current I_C^d. Since the ratio is temperature-dependent, which comes from the temperature dependence of the critical current I_C, the temperature dependency of I_C is studied separately. In order to evaluate the critical current at different temperatures T, a current-voltage measurement is executed, which is repeated for different temperatures between 4.2 K and T_C with several 4 nm thick detectors. Figure 4.16 depicts the extracted I_C for sample 1 of Table 4.1. For theoretical comparison, the analytical current I_C^d of the temperature dependence of the depairing critical current is also plotted as red line. I_C^d is given by

$$I_C^d = \frac{4\sqrt{\pi}(e^\gamma)^2}{2 \cdot l\zeta(3)\sqrt{3}} \cdot \frac{w\Delta(0)^2}{eR_s\sqrt{D\hbar k_B T_C}} \left[1 - \left(\frac{T}{T_C} \right)^2 \right]^{3/2} \tag{4.7}$$

which is suggested by Bardeen [91]. $\Delta(0)$ is the superconducting gap at zero temperature, $\zeta(3) = 1.202$ is the Zeta-function, $\gamma = 0.577$ is the Euler's constant, $D \approx 0.5$ cm^{2}s^{-1}

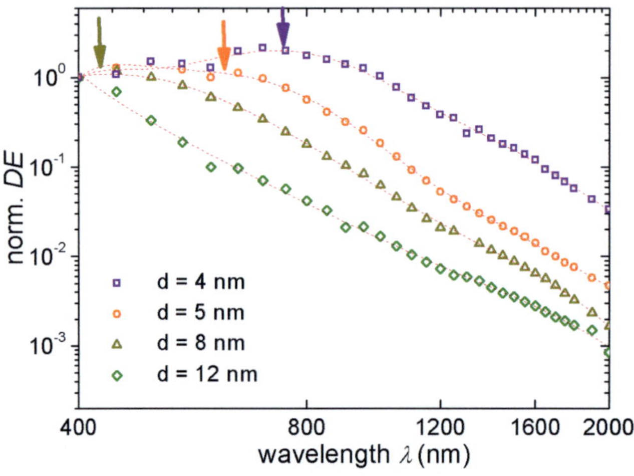

Figure 4.15: Dependence of the detection efficiency on the film thickness. The *DE* is normalized to its value at $\lambda = 400$ nm. The *DE* for all meanders is measured at $I_B = 0.95 \cdot I_C$ and $T = 6$ K. The arrows mark the cutoff wavelengths. [HRI$^+$10b]

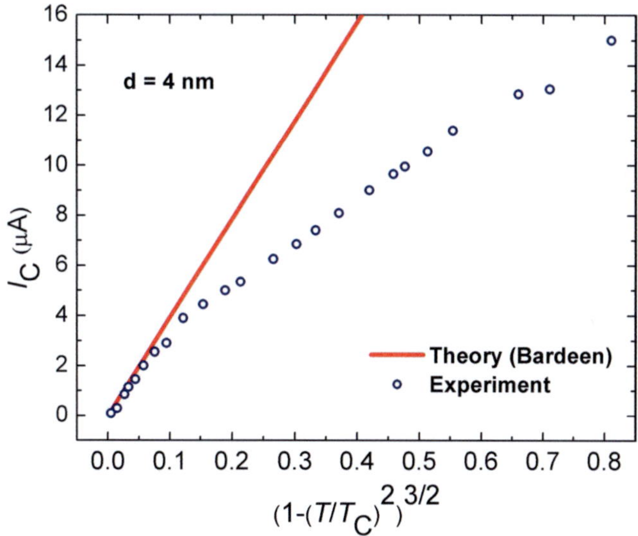

Figure 4.16: Experimental critical current I_C compared to the analytical critical current I_C^d based on Bardeen's depairing critical current of a 4 nm thick meander. [HRI$^+$10b]

is the typical electron diffusivity in our NbN films [39], and e is the electron charge. This representation shows explicitly the dependence of the depairing critical current on the energy gap $\Delta(0)$. At $T \approx T_C$ in the limiting case of weak coupling $\Delta(0) = \pi e^{-\gamma} k_B T_C$, Eq. 4.7 can be simplified to the standard Ginzburg-Landau form. To define I_C^d for the meanders, the adapted energy gap $\Delta(0) = 2.07 \cdot k_B T_C$ measured in thin NbN films is used [HDH$^+$12], [92] and [93]. The I_C of all meanders follow the course which is depicted in Fig. 4.16. At $T \approx T_C$ they coincide with the depairing critical currents I_C^d and increase to approximately $0.5 \cdot I_C^d$ at temperatures less than one half of the critical temperature T_C. One has to respect, as well, that in the extremely dirty limit ($l \ll \xi_0$), which is valid for the films [39], Bardeen's depairing current at $T \approx 0.5 \cdot T_C$ amounts to only 85 % of the numerically computed depairing current [94]. An analytical estimation of the cutoff wavelength λ_0 therefore requires a correction.

Using Eqs. 4.7 and 4.6 and respecting the difference between Bardeen's critical current and the true numerical depairing critical current [94], the cutoff wavelengths for the bias current $I_B = 0.95 \cdot I_C$ are computed. The nominal line widths w and the square resistances R_S are taken from Table 4.1 and the standard Bardeen-Cooper-Schrieffer temperature dependence of the energy gap is used [18]. The results are depicted in Table 4.1 and the interesting dependence λ_0 on R_S is shown in Fig. 4.17, as well. The best agreement with the experimentally extracted cutoffs is obtained for the effectiveness parameter $\varsigma = 0.43$, which fits to already reported values [21], [34]. The most critical factor in this calculation is the poorly defined density of Cooper pairs n_s, since it directly influences the effectiveness

ς. Nevertheless, there is a good agreement between the cutoff extracted from the measurements of the detection efficiency depending on I_B and the film thickness d, respectively the square resistance R_S and the theoretical model.

The non-dropping DE beyond the cutoff is a contradiction to the hot-spot mechanism, which indicates that there should be another mechanism for the appearance of the normal-conducting belt. Based on the hot-spot model the photon detection beyond the cutoff could be explained by constrictions [95], which occur with a certain distribution along the wire. They may appear with either geometrically reduced cross-sections or reduced superconducting gap, which directly influences the parameters of the cutoff condition and leads to a certain smoothing of the cutoff. A photon absorption may lead to a normalconducting belt close to these constrictions, even, if the photon energy is not strong enough to initiate a suitable large hot-spot at regions with nominal parameters far away from these constrictions. In comparison to the exponential decay in the DE over more than two decades in the photon energy, this explanation would require at least some constrictions with cross-sections between one half and twice the mean value of the line width. Such geometrical constrictions should have the lengths larger than the coherence length ξ and hence could be visible in an electron beam microscopy picture. Additionally, such constrictions should have a measurable influence on the critical current I_C of the wire in relation to the depairing critical current I_C^d over the full temperature range. Careful inspection with SEM, transmission electron microscopy, and atomic force microscopy [39], [47] show a spread of the geometrical parameters of less than 10 % in the cross-section of the meander lines.
An alternative reason for constrictions in the cross-section, non-homogeneity of more than

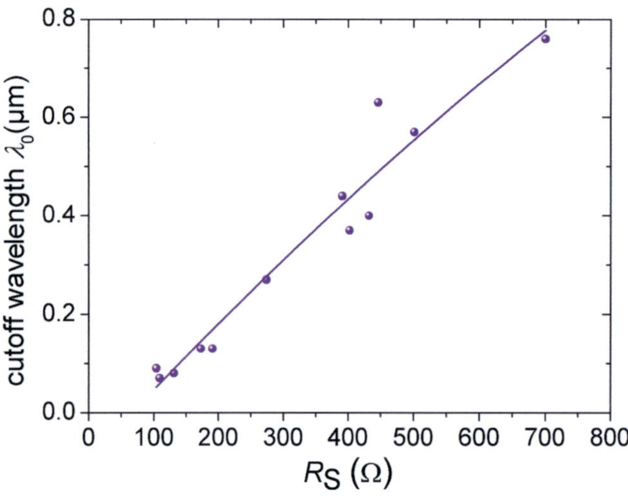

Figure 4.17: Dependence of the cutoff wavelength λ_0 on R_s calculated from the film and operation parameters.

75 % in the energy gap along the meander line could explain the exponential decrease in the *DE* over two decades in wavelengths while a non-homogeneity of only 25 % would already explain the reduction in the measured critical current I_C. However, a depiction of the energy gap in the films, which is performed by means of scanning tunneling microscopy, has revealed a non-homogeneity of less than 15 % based on a lateral resolution comparable to the coherence length [93]. Moreover, it is quite sure that R_s has no systematic dependence on the line width [47] which can not be explained by experimental inaccuracies. One can conclude that the mean size of possible grains should be at least one order of magnitude less than the line width. Consequently, the size of the non-equilibrium hot spot $\approx \sqrt{D\tau}$ is larger than the granularity of the material. Therefore, film granularity can not be the reason for a local variation in the *DE* leading to two orders of magnitude spread in the count rate but may be a reason for a certain inhomogeneity in the current distribution in the meander which may explain in part the deviation of the experimental critical current I_C from I_C^d.

The mean amplitude A_0 of the voltage transients, which the detector returns in response to single photons, changes with the increase in the wavelength (see Fig. 4.11, [20]). The change of the mean amplitudes seem to start from the same wavelength range as the cutoff. That means the change of A_0 has a certain dependence on the wavelength but is not proportional to the photon energy. Moreover, even if one assumes that the absorbed photon energy defines the hot spot which might vary the resistance at the beginning of the appearance of the normalconducting belt, one would expect for the shorter wavelengths a higher resistance and therefore, a higher mean amplitude of the detector response pulses. The contrary situation is the case.

Another point, small constrictions in the range of 10 to 20% of the meander wire (as they can be seen by SEM imaging) could influence the amplitude of the transients, as well. However, the amplitude change seems to follow more likely a course which is not defined by a random geometric constriction in the meander width [96]. The open question stays, if the amplitude A_0 of the transients is only controlled by the evolution of the normalconducting belt at the second stage of the detection scenario and unconnected at all to the physical mechanism, which initiates the appearance of the resistive belt. However, the fact that well in the cutoff regime photon counts have the same amplitudes as dark counts likely indicates that they originate from the same physical process deviating from the hot-spot detection process [20].

The previous discussion does not mean that constrictions and non-homogeneity have no impact to the physical description of the device response but their actual amount and strength are not sufficient for quantitative evaluation of the available experimental data.

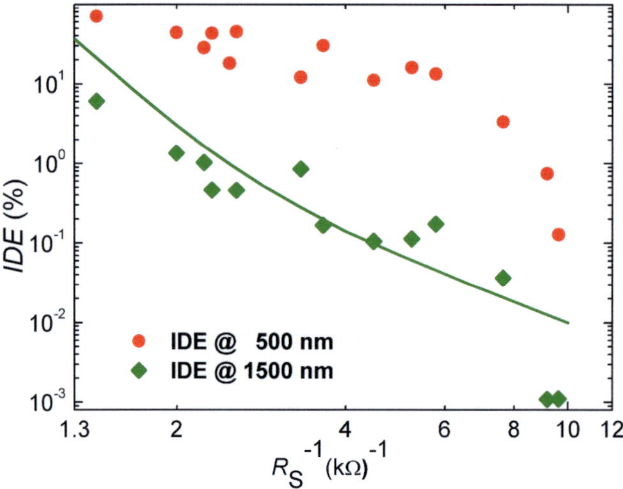

Figure 4.18: Intrinsic *DE* for the meanders with different thicknesses at the wavelengths 500 and 1500 nm. The *IDE* is plotted vs reciprocal square resistance R_S. The solid line is the computed *IDE* for vortex assisted photon detection [HRI$^+$10b].

4.1.5 Intrinsic detection efficiency for different film thicknesses

In Fig. 4.18 the intrinsic detection efficiencies of the detectors are depicted versus their reciprocal resistance per square R_S^{-1}. The experiment is actually focused on the dependence of the detection efficiency on film thickness, but the comparison to R_S allows an increase in the accuracy of the analysis. Using the pure film thickness d would require knowledge of the specific resistivity ρ, which itself depends on d for thin films [39].

One can see in Fig. 4.18 that for $\lambda = 500$ nm the *IDE* varies little until d reaches 7 nm $(R_S^{-1} \approx 4\,\mathrm{k\Omega^{-1}})$ and drops quickly with the further increase in the thickness. *IDE* reaches almost 70 % for the thinnest films and drops to about 20 % for the meander from the 9 nm $(R_S^{-1} \approx 5.7\,\mathrm{k\Omega^{-1}})$ thick film. A different situation is given at the wavelength 1500 nm. Here, *IDE* monotonically decreases with the decrease in R_S. The two different behaviors depending on wavelength enforce again the assumption that two different mechanisms define the photon detection.

4.1.6 Vortex-based detection mechanism for infrared radiation

The earlier experiments clarified that an additional mechanism is required to explain the detection above the spectral cutoff. One possible extension to the hot-spot detection mechanism, introduced by [97] and [98], is given by the idea of thermally excited vortices, which cross, activated by the photon energy, with a certain probability the meander stripe, forced by the Lorentz force and produce a voltage transient on the meander. A vortex consists of screening currents around a magnetic flux quantum. It forms a flexible tube in the superconducting film, where no Cooper pairs can enter. If the criterion $4.4\xi(T) \leq w$ is fulfilled as in our situation, vortices can enter a superconducting line [99], [100]. $\xi(T)$ is the superconducting coherence length. A typical value of $\xi(T)$ in the NbN films is 4 nm at zero temperature [47].

The crossing of the magnetic vortex in the superconducting stripe produces a screening current with a local current density j around the magnetic core (see Fig. 4.19). The circular screening current is expended by a force. That means, the self-energy of the vortex needs to be high enough to apply this force to enable a vortex nucleation. The self-energy is given from the integral of the magnetic force with regard to the boundary conditions like the meander width w [98]:

$$U_{pot}(x) = -\frac{\phi_0^2}{2\mu\Lambda\pi} \cdot ln\left(\frac{w}{\pi\xi}sin\left(\frac{\pi x}{w}\right)\right) - \frac{I_B\phi_0 x}{w} \tag{4.8}$$

Here, Φ_0 is the flux quantum and $\Lambda = 2\lambda_L^2/d = 2\hbar R_S/(\pi\mu_0\Delta)$ defines the Pearl length [101], the effective penetration depth for very thin films, with λ_L and d being the magnetic penetration depth and the thickness of the film. x is the position between the edges of the superconducting line. In Eq. 4.8 the first term describes the self energy of the vortex along the width of the line, which is reduced by the Lorentz force of an applied bias current I_B described by the second term. The conditions of the vortex nucleation are not defined by the potential barrier $U_{pot}(x)$ along the width w but the maximum of this barrier under a given current I_B. A further influence of the barrier due to the earth magnetic field can be neglected in this model [98].

The maximum height of this barrier can be calculated by algebraic definition of the maximum turning point of Eq. 4.8. (transformation by A. Semenov [HRI$^+$10b]):

$$U = \varepsilon_0 \left\{ ln\left[\frac{w}{\pi\alpha\xi} \cdot \frac{1}{\sqrt{1+\left(\frac{I_B}{I_0}\right)^2}}\right] - \frac{I_B}{I_0}\left[\tan^{-1}\left(\frac{I_0}{I_B}\right) - \frac{\pi\alpha\xi}{w}\right] \right\} \tag{4.9}$$

The definitions for the current and the energy factor are $I_0 = \Phi_0/(2\mu_0\Lambda)$ and $\varepsilon_0 = \Phi_0^2/(2\pi\mu_0\Lambda)$. With the quantum resistance, which is $R_0 = \frac{h}{2e^2} \approx 13$ kΩ, one can approximate the energy dimension $\varepsilon_0 = (\pi/4) \times (R_0/R_S)\Lambda$. At $T \approx T_C$ the current dimension attribute to the depairing current $I_0 = 4.08\, I_C^d \cdot \xi/w$. Using this relation at typical operation temperatures $T \approx 0.5 \cdot T_C$ and $I_B/I_C^d \approx 0.4$, a current ratio I_B/I_0 in Eq. 4.9 close to one is reached. For a fixed operation point I_B/I_C^d, the both parts in square brackets in Eq. 4.9 can be considered as the geometric factor, however with no influence of the thickness. The constant $\alpha = 4/\pi$ is required to describe the smallest distance (in units of ξ) between the edge of the line and the position where the vortex can nucleate [102] and develops its screening current. This distance is not required to define the shape of the barrier [103] but is important for its absolute height.

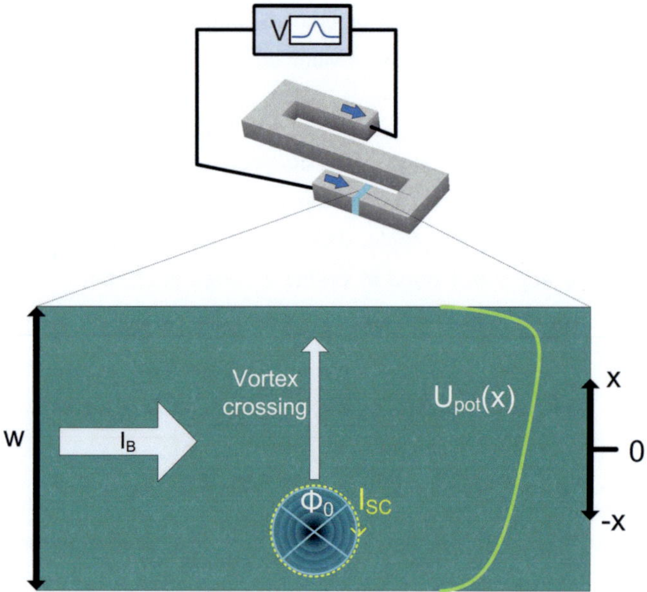

Figure 4.19: Scheme of the vortex penetration. The potential barrier is symbolically depicted for an arbitrary bias current (compare [98]).

Although this barrier exists, there nevertheless is a certain thermodynamic probability P that the vortex crosses the barrier and enters into the meander. After nucleation, the vortex will be accelerated by the Lorentz force and moved to the opposite edge of the meander. This leads to a count event since the moving vortex causes energy dissipation.

The conclusion of this mechanism is that even beyond the cutoff an absorbed photon locally increases the probability P for the vortex to nucleate at the edge of the line and to transit through the meander. The nucleation probability P is proportional to the bias current I_B and thermodynamic excitation rate $P \propto I_B \, exp(-U/k_B T)$ [97]. An absorbed photon produces a small energy impact in the film, which locally reduces the energy gap by the amount $\delta\Delta$. In the same way the height of the barrier is reduced by the amount δU. So, δU is proportional to the photon energy, as long $\delta\Delta \ll \Delta$. Under these conditions the count rate in response to the absorbed photons and consequently the *IDE* should then qualitatively follow the vortex excitation probability δP.

$$\delta P = \Omega \cdot I_B \cdot e^{(-U/k_B T)}[e^{\delta U/k_B T} - 1] \tag{4.10}$$

Ω [1/A] is a coefficient which defines the attempt rate of vortex penetration attempts. This is crucial since it is supposed that the number of attempts of a vortex penetration is much larger than the probability of an absorption of a photon itself. The term in square brackets gives the *DE* a exponential dependence on the absorbed photon energy. δU should be almost independent of the film thickness because for thin films the local reduction in the energy gap Δ scales with the reciprocal film thickness [104]. Moreover, the height of the potential barrier U scales approximately with $1/R_S$ which means an exponential decrease in the *IDE* with increasing thickness d, which also should be available in the *DE*. Since in our operation conditions it is valid $U \ll k_B T$, *DE* as function of the photon energy could only saturate when $\delta U = U$. Based on Eqs. 4.9 and 4.10 this basically happens when the energy gap $\Delta = 0$. It follows that the photon solely must locally destroy the superconductivity in the meander volume. Indeed, applying a bias current, even smaller photon energy suffice to satisfy the criterion for the evolution of the hot-spot detection (Eq. 4.6).

We can sum up that according to the two detection mechanisms, the *DE* should be constant at short wavelengths and small thicknesses and should have a quasi-exponential decay with the further increase in both the thickness or the wavelength. The experimental data qualitatively agree with this description and are also consistent with earlier obtained results for different NbN films [38], [105], [104]. In Fig. 4.18, at $\lambda = 1500$ nm, we are in the region, where the thermal excitation of magnetic vortices is supposed to realize photon detection. Here, the *IDE* decreases quasi-exponentially with increasing film thickness. Contrary, for 500 nm photons, the hot-spot response dominates up to a certain thickness and the *IDE* is saturated at almost constant values. However, for the thicker films the detection mechanism changes at the energy of 500 nm photons.The efficiency shows a quick drop with the increase in the thickness.

For testing the vortex model quantitatively, the relative decrease in the Cooper-pair density with the relative decrease in the energy gap $\delta n_S/n_S = \delta\Delta/\Delta$ is compared in co-work with A. Semenov. Applying to Eq. 4.6 the best fit value $\varsigma = 0.43$, one further gets the relative reduction of the energy gap as function of the photon energy. The behavior of the barrier height for the vortex entry is calculated with $\delta U = U(\Delta) - U(\Delta - \delta\Delta)$ with the constant ratio I_C/I_C^d. The consequential *DE* is calculated using Eq. 4.10. Here, the width (w = 90 nm), the operation temperature (6 K) and the transition temperature T_C (13 K) are set fixed and the coefficient Ω and the ratio I_C/I_C^d are used as fitting parameters. The best agreement with measurements is obtained with $I_C/I_C^d = 0.41$, which is only slightly smaller than the experimentally defined value for the 4 nm film, and an attempt rate $\Omega = 8 \times 10^6$ A^{-1} for $I_B = 0.95 \cdot I_C$. Ω has to rise up to fit the experimental data at smaller bias currents.

The fitting the probability of Eq. 4.10 to the measured *DE* is depicted in Figs. 4.20, 4.21 and also on the *IDE* in Fig. 4.18 with solid lines. The model describes quite well the dependence of the *DE* on the experimental wavelength and its decay with the film thickness d for thicknesses up to 10 nm. The detection efficiencies of the two 12 nm thick meanders are remarkably lower than the predictions of the model. This deviation could be explained in this way that the thick films do not fully obey a two-dimensional electron diffusion, which the current model assumes. The wavelength of 1500 nm moreover has a larger spread of data points than the 500 nm. This could be caused by a spread of the critical temperatures

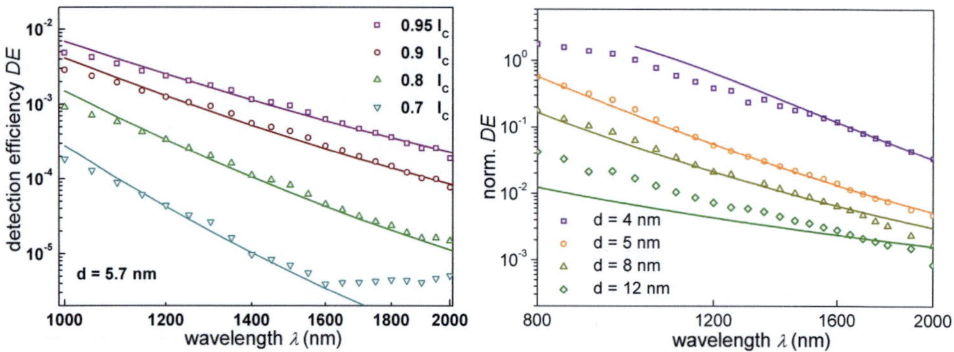

Figure 4.20: Vortex model fitted on the *DE* measurements over wavelength for different I_B.

Figure 4.21: Vortex model fitted on the *DE* measurements over wavelength for different thicknesses d.

T_C and the line widths w and also by the fact that the nucleation of vortices is strongly influenced by the geometric quality of the edges of the meander line. It has been reported by many groups [96], [106] that the performance of an SNSPD improves at lower temperatures which contrary explains the relatively low infrared *DE* obtained in this work at 6 K. However, on the other side neither hot-spot nor vortex detection mechanism predicts the rise in the *IDE* with decreasing operation temperature T. A local collapse of the superconductivity always becomes less probable at low temperatures due to the enlarged condensation energy, which leads to a higher superconducting gap Δ and a decrease in the rate of any kind of thermal excitations. This description is part of all fluctuation models with activation energies. Mainly the unknown dependence of the attempt rate on current and temperature seems to be the explanation for this apparent contradiction in context with infrared photon detection. It has been demonstrated that the attempt rate has to be increased at low temperatures to explain the absolute values of the dark count rate [47], [45]. A further explanation is analyzed in [EIS$^+$13], which discusses a temperature dependent diffusivity $D(T)$. Another crucial note is that the current and temperature dependence of the *DE* may also be caused by local differences in the thermal coupling between the film and the substrate [107]. These variations determine locally whether and how the normalconducting belt will rise or shrink in this local part of the meander line although such inhomogeneities do not change the critical current or the critical temperature of the meander line. For any bias current I_B, there may exist sections, which are better thermally coupled with the substrate and where the domain cannot grow. Increasing the bias current I_B at a constant temperature or decreasing the operation temperature at simultaneously constant relative bias current may activate some of such sections. The respective increase in the *DE* can be formally accounted for as the temperature- and current-dependent attempt rate. Meanwhile, there exist extended theoretical descriptions, which discuss more in detail the vortex supported detection and can help to improve the understanding of infrared detection [108] and [109].

The analysis of the detection efficiency has shown that in SNSPDs the absorption of the meander line influences noticeably the value of the *DE* as well as its spectral dependence. Extracting the intrinsic detection efficiency for meanders with different thicknesses, it was found that the intrinsic detection efficiency reaches a level up to 70% and remains almost independent of the wavelength and thickness up to the limit, which is defined by the cutoff criteria of the hot-spot detection mechanism. At larger wavelengths, an alternative mechanism based on the excitation of magnetic vortices over the potential edge barrier describes well variation in the intrinsic detection efficiency with both, the wavelength and the thickness. Moreover, it is shown that this alternative detection mechanism influences the mean amplitude and amplitude distribution of an SNSPD response.

4.2 Modeling of the dark count rate vs. thermal coupling, temperature and stoichiometry

Dark counts can be considered as the intrinsic noise of an SNSPD. A very low dark count rate is a preferable parameter of an SNSPD system. In this subchapter the focus is set on the origin of dark counts as result of thermally activated vortex crossing the meander line and on a discussion about possible methods to reduce the dark count rate [HRI$^+$12], [HDH$^+$12].

4.2.1 Dark count model

In section 2.2.2, ideas about the origin of the dark counts in an superconducting nanowire with widths in the range of $w = 100$ nm are explained. According to [47] the thermal excitation of vortices in a meander stripe most likely describes the dark count rate behavior. A certain influence of the other models cannot excluded on the other side. The thermal excitation model assumes that dark counts are caused by the same mechanism, which is responsible for infrared detection based on thermally activated vortex distribution [48], [47], [HRI$^+$10b] because a fluctuation in the meander line has not enough energy to directly produce a hot spot and dark counts have the same pulse shape concerning the amplitude distribution than the response pulses of infrared photons [104].

Vortices in superconductors require a certain size, which means that their existence is limited to a certain geometry of the superconductor. According to Likharev, vortices can exist in a small line if the width is larger than 4.4ξ, where ξ is the coherence length [99]. If this condition is fulfilled, thermal activated vortices can nucleate from geometric consideration at the edges of the line.

According to section 4.1.6, a vortex nucleation at the edge of a nano stripe is hampered by a potential barrier U defined by the Gibb's free energy of the screening current of the magnetic flux, which is contained in the vortex. If the vortex overcomes U, the crossing is forced by the action of the Lorentz force, which is imposed by the bias current I_B. The movement of the vortex induces a voltage pulse on the detector, and the dissipation in the line leads to a normalconducting belt. At this moment the narrow line behaves in the same way as in the case of photon detection: The normalconducting belt causes a voltage transient at the readout line [48], [47].

Since, in the typical operation point, the bias current I_B approaches the critical current I_C, the detector is very close to the phase transition. At these conditions only a small activation energy ΔE is required to drive a portion of the nanowire to the normal conducting state. However, according to [98] typical bias currents of SNSPDs are too low to decrease the superconducting gap Δ and therefore, the potential barrier U to a level in the range of $k_B T$, which would lead to a sufficient high probability of thermal activation of vortex crossings. In [47] a k_B-normalized potential U of ≈ 300 K at I_B close to I_C is calculated for a typical meander stripe. Nevertheless, there is a measurable probability $P \propto \Omega \cdot I_B \cdot e^{\frac{\delta E}{k_B T}}$, that a single vortex crosses the stripe, since the probability P of a crossing correlates with the attempt rate Ω, as well [97]. A high Ω enables even for lower $U/k_B T$ ratio a certain number of vortex crossings. Moreover, one can assume that the theoretical maximum of the barrier U is not reached, if one takes the latest results from [49] into account. They demonstrate that the dark counts originate mainly at the turning points of the meander lines, where a certain current crowding exist and certain constrictions in the line may appear.

As conclusion from the last statement, a possible way of reduction of the DCR consequently is to fabricate a more uniform detection line, what is not the focus of this work but can be found e.g. in [HRD$^+$13].

In general, reducing energy fluctuations is one of the possible ways to decrease the DCR, under the condition that it is clear, where they originate. One can assume that noise or thermal impact along the supply and readout lines superpose the bias current. Improved cooling of the bias and readout cable could be a way, reduction of noise and distortion in the bias path by using filters and low noise sources is possible, as well. However, at a certain point, a limitation of the optimization is given because the basic thermal fluctuation energy $k_B T$ at the detector position still exists, which always enables a certain number of vortex crossings.

Based on the model of vortex supported dark count evolution, several alternative approaches are analyzed to decrease the dark count rate which manipulate Ω, U and T in Eq. 4.10:

- Reduction of vortex fluctuations in the superconductor by improved cooling interface of the detector mounting.

- Variation of the excitation energy of a dark count by different operation temperature conditions.

- Enhancement of the potential vortex barrier of the meander line by variation of film parameter.

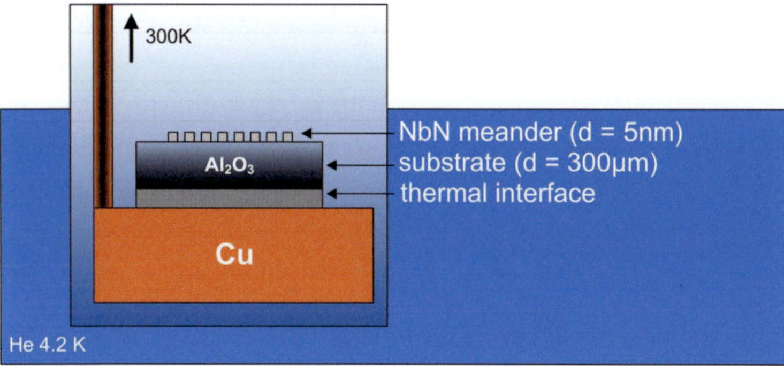

Figure 4.22: Scheme of thermal coupling of an SNSPD. The detector is mounted with the full bottom plane of the substrate to a copper plate. The detector is illuminated from above.

A reduction of the bias current I_B also reduces the dark count rate because both, the superconducting gap Δ is increased and the Lorentz force, which is proportional to I_B, is reduced, which leads to an increased potential barrier U and the attempt rate Ω of the vortex penetration is reduced. However, the detection efficiency decreases significantly, when the ratio I_B/I_C drops just a few percent below 1. Hence, a reduction of I_B is not a convenient way to decrease the dark count rate.

4.2.2 Reduction of vortex fluctuations in the superconductor by improved cooling interface of the detector mounting.

In this subsection the dark count rate versus the quality of the cooling interface between detector chip and detector block is analyzed. The section bases on the published work in [HRI$^+$12].

Dark counts are activated in dependence on the relation between the thermal energy $k_B T$, defined by the mean operation temperature T, and the potential barrier U of the meander line. On the other side, the attempt rate Ω is a variable parameter, as well. The question is, in which way thermal fluctuations influence the thermal conditions in the superconductor. Does a fluctuation produce a short local temperature peak in the film or a temporally broad but moderate increase in the temperature [110].

A possible explanation is given looking at the thermal capacity c_{system} of the complete detector system. An SNSPD can simple be considered as a two-layer system consisting of the NbN layer, and the simply substrate layer. The SNSPD itself is mounted to a copper holder. The mounting material acts as an interface between the detector and the copper volume (see Fig. 4.22). The mounting material has a certain thermal conductivity k_w. Comparing the two volumes directly, one can assume that the copper block has a very inertial reaction of the temperature to an energy input and the pure SNSPS will react quite more sensitive. The cooling of the NbN SNSPD occurs via the sapphire substrate. The interface between the copper plate and the detector chip strongly influences the common thermal capacity c_{system}.

The better the detector is coupled to the copper plate the less strongly causes a fluctuation a possible temperature peak in the film. The aim of this experiment is to analyze the *DCR* using interfaces with different thermal conductivity k_w. k_w is varied by using different materials for mounting the detector to the copper plate. The change of k_w during the cooling process is neglected in this consideration because the relation in volume between the detector and the copper block is the same at low temperatures and this approach investigates a relative improvement of the common thermal capacity and not an absolute change of c_{system}.

For this experiment, an SNSPD is used, which consists of a 5 nm thick magnetron sputtered NbN film on one-side polished sapphire. The meander structure with a width of 97 nm is made by electron-beam lithography and argon-ion milling. It has a critical temperature $T_C = 12.4$ K and a critical current $I_C = 42$ μA (@ 4.2 K). For the measurement the cryogenic dip stick from chapter 3 is used. The output signal is amplified by three rf amplifiers with an effective bandwidth between 50 MHz and 1.7 GHz. The fiber input can be completely covered for dark count measurements. The detector is mounted on the massive copper holder. The physical temperature of the detector in this experiment is defined via the critical current I_C of the meander line. The repetitive measurement of I_C allows to check the stability of the mean temperature T of the system during the experiment. Small temperature fluctuations δT in the film are not resolvable with this method. However, one can make sure that the mean temperature T is nearly kept constant for all measurements.
The following three cases are studied:

1. The SNSPD is pressed against the cold plate with silicon grease in between and fixed on the edge with superglue.

2. The SNSPD is glued to the cold plate with little conductive silver paste.

3. The SNSPD is glued to the cold plate with a good amount of conductive silver paste.

For all three methods the *DCR* is measured versus I_B. Silicon grease has a much lower thermal conductivity ($k_{w,\,sg} = 10$ mW/mK at 4.2 K [111]) than conductive silver ($k_{w,\,cs}$ – depending on impurity, for a $RRR_{silver} = 30$, $k_{w,\,cs} = 300$ W/mK at 4.2 K [112]). The thermal conductivities of these two interfaces vary by more than 5 orders of magnitude.
Figure 4.23 shows for all studied interfaces the typical exponential increase of the dark count rate with linear increase in bias current I_B. The main difference is the onset current and the maximum value. A background count rate of 1 s^{-1} is generated by the electronic noise in the setup. The dark counts of the interface with silicon grease start at the bias current $I_B = 0.94 \cdot I_C$, which is less than the typical operational bias current $I_B = 0.95 \cdot I_C$. Both interfaces, based on silver paste, lead to a *DCR*, which appears at currents approximately $0.97 \cdot I_C$. Consequently, at typical I_B the *DCR* is < 1 s^{-1}.
The maximum *DCR*, which is measured for I_B close to I_C, differs within two orders of

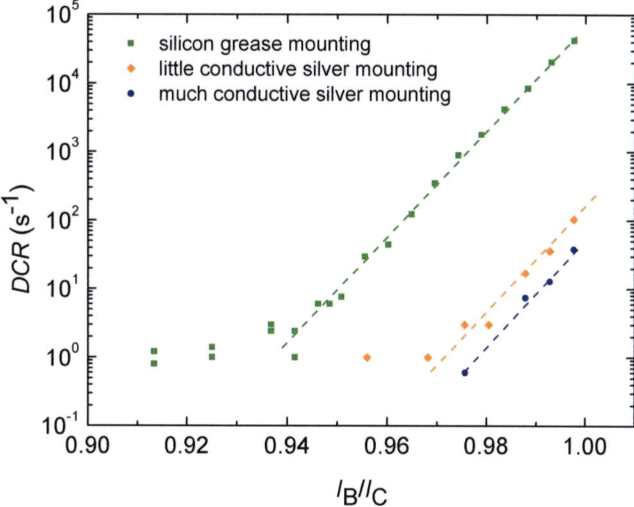

Figure 4.23: Dark count rate versus I_B/I_C depending on thermal interface between detector chip and cooling plate. The dark count rate is measured for three different types of detector mounting.[HRI$^+$12]

magnitude between the three types of detector mount. The variant, which uses only little conductive silver paste, hasn't a completely filled contact area between the substrate and the cold plate. Consequently, the thermal conductivity $k_{w,s}$ is non-uniform over the contact area. Obviously, this non-uniform thermal flow results in a slight increase of the *DCR* for all bias currents compared to the interface with laminar filled conductive silver paste. For both interfaces with silver paste, the *DCR* is less than 100 s^{-1} even at I_B close to I_C.

Considering the difference of factor 10,000 of the thermal conductivity between the grease and the silver paste, one can conclude that the conductive silver paste indeed has a better thermal interface to the massive copper plate. Energy fluctuations, which are always present in the system, have less impact to a local temperature increase in the massive copper plate and therefore, also less influence to the detector, when it is well coupled to the plate. Silicon grease provides worse coupling between the detector and the copper plate. Therefore, thermal fluctuations in the detector might enhance the probability of a vortex crossing and cause a larger *DCR*. Using silver paste for mounting, the dark counts can be almost neglected in the typical operation range of $I_B = 0.9$ to $0.95 \cdot I_C$.

The experimental results are compared to the predictions of the vortex model. Since the change of the thermal conductivity can not be defined in exact numbers, the comparison is made qualitatively only. In Fig. 4.23 all three measured dark count rates are fitted with an exponential fit. All *DCR* change exponentially with I_B, which comes from the direct change of the superconducting gap Δ in U with I_B. However, for all curves the denominator in the exponent is the same, which can be seen in the parallel fits. This means that the exponential part of the probability in Eq. 4.10 is not affected by the change of the thermal conductivity of the interface. Therefore, the vortex barrier is unchanged for all types of thermal conductivities. Contrary, the improvement of the thermal conductivity k_w reduces the attempt rate Ω of vortex crossing attempts, which has a linear dependence in Eq. 4.10. It can be seen by the higher I_B value of the start of *DCR*. The linear dependence demonstrates, as well that the effect is not correlated to a change of temperature T. This would claim an exponential dependence of *DCR* on the variation of the thermal interface. The discussion about the impact of fluctuation on the temperature could be expanded by the latest results on behavior of vortices in superconducting films [113], which have shown that dendritic magnetic flux avalanches are strongly correlated to thermal-magnetic instability. One can assume from this results that this could be the effect, which changes the attempt rate of vortex nucleation in case of dark counts, as well. This analysis considers the homogeneity of the thermal coupling between superconducting film and substrate at constant temperatures versus the influence on vortex behavior.

One can conclude that a variation in the thermal interface is able to tremendously reduce the *DCR*. However, it is not an effect of change of the mean temperature in the detector system. Since detector parameters like detection efficiency are dependent on the mean temperature of the system [EIS$^+$13], the detection efficiency should be independent of the improvement of the interface.

4.2.3 Variation of the excitation energy by different operation temperature conditions

In this section, the dark count rate versus the operation temperature T is analyzed. Comparing the measurements from different research groups, one finds out that typical operation temperatures of SNSPDs are in the range between 300 mK and 4.2 K. An analysis of the dark count rate for this range can be found in [114]. They show, that *DCR* reduces with decreasing operation temperature. With the IMS dip stick it is possible to realize measurements from 4.2 K to T_C. This broader range allows an evaluation of the dark counts more in detail for better understanding of the different dark count models.
The experiment is realized with a 4 nm thick detector. The detailed data of this sample are:

Table 4.2: Detector parameter, *DCR* over temperature

$d\,(\mathrm{nm})$	$R_S(\Omega)$	$T_C(\mathrm{K})$	$j_C^m(\mathrm{MA/cm^2})$	$2\cdot\Delta(\mathrm{meV})$
4.2	430	12.5	4.6	4.4

The experiment is executed in the cryogenic dip stick as described in chapter 3. The fiber input is completely blocked for dark count measurements. The operation temperature is measured by a Lakeshore temperature diode in the copper block directly below the detector.

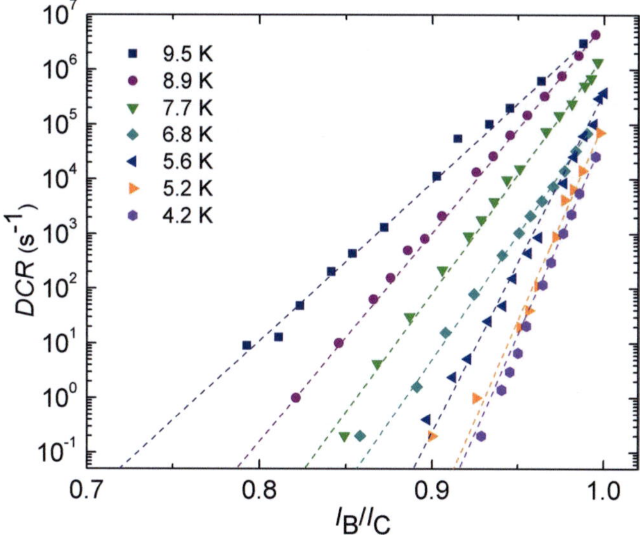

Figure 4.24: Dependence of the dark count rate versus I_B/I_C on temperature. The dashed lines correspond to a exponential fit for each measurement series.

The temperature variation is realized by changing the distance between the sample and the liquid helium level and a heat resistance. Therefore, an adequate time period has always to be waited to get the temperature around the detector in an constant thermal equilibrium. The measurements are done in part by the author and Dagmar Henrich.

In Fig. 4.24, the bias current I_B is normalized to the critical current I_C, due to the fact that I_C varies with the temperature T. One can see several aspects. The maximum DCR increases with increasing temperature. At 4.2 K, DCR is almost three orders of magnitude smaller than at the highest measured operation temperature (9.5K). Moreover, DCR arises already at $I_B < 0.8 \cdot I_C$ close to T_C.

The measurement series stops at 9.5 K, which is still below T_C (12.5 K). At this temperatures, the critical current I_C is already at the limit to produce pulses with sufficient SNR, to resolve the pulses from the background noise.

The variation of temperature T is compared to the vortex model, as well. Again, each dark count curve is fitted by an exponential function as defined by the model in Eq. 4.10 with a rational function in the exponent. The denominator of each rational function is extracted and plotted versus the operation temperature T (see Fig. 4.25). According to Eqs. 4.10 and 4.11, the temperature T defines the thermal fluctuations $k_B T$ and the superconducting gap Δ. The potential barrier U is proportional to $1/\Delta$, since $\varepsilon_0 \propto U$ (Eq. 4.11) and ε_0 can be written as:

$$\varepsilon_0 = \Phi_0^2/(2\pi\mu_0\Lambda) \quad \text{with} \quad \Lambda = 2\lambda^2/d = 2\hbar R_s/(\pi\mu_0\Delta) \tag{4.11}$$

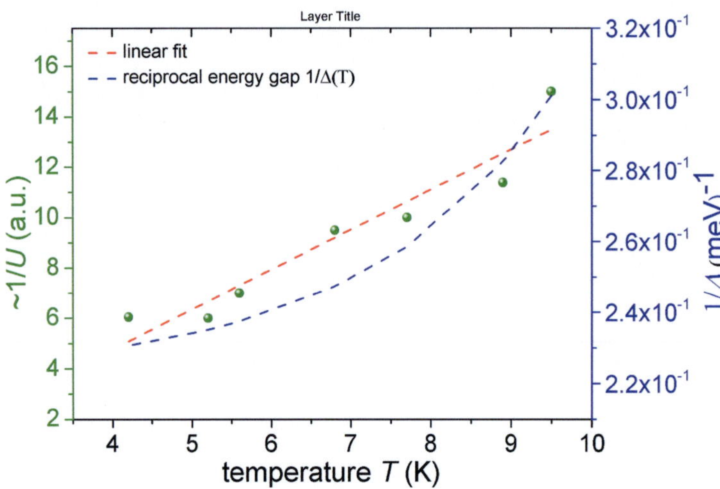

Figure 4.25: Extracted denominators of the rational functions in the exponential fits of *DCR* over *T* from the curves in Fig. 4.24.

T and $\Delta(T)$ are consequently part of the denominator of the exponent of the probability function Eq. 4.10. For this reason, a linear fit is set to the extracted denominators in Fig. 4.25 and the reciprocal $\Delta(T)$ according to Eq. 4.2 is plotted, as well. One can see, that the denominator follows mainly the linear dependence as expected from the $k_B T$ term, the closer the temperature comes to T_C, the more one could assume that the shrinking superconducting gap Δ influences the course of the denominator in accordance with the vortex model, as well.

The temperature T influences the attempt rate Ω, as well [47]. Ω changes the starting bias current of the dark counts. However, this point is defined by the exponential function, as well. Therefore, it can not be separately evaluated.

One can conclude that over the complete range of temperatures, the dark count rate decreases with a decrease of the temperature T. T directly affects the vortex barrier and controls in this way the probability of vortex hopping. T mainly influences the thermal energy $k_B T$. However, the dependence of $\Delta(T)$ on T also influences the DCR close to T_C.

4.2.4 Enhancement of the potential vortex barrier of the meander line by variation of film parameter

In this section, the dark count rate versus the film composition is analyzed. In section 4.1.2, the standard process of film deposition is described. The films, which are used in the previous experiments, are optimized for highest T_C and I_C. This optimization can be reached by finding the optimal sputter current I_{SP}, which defines during deposition the stoichiometry between Niobium and Nitrogen of the final film. The details are described in [HDH$^+$12]. In this work Henrich et al. have analyzed in which way, a change of the composition improves the detection efficiency.

Again taking the vortex model, the parameter which mainly changes the potential barrier $U(\varepsilon_0)$ (see Eqs. 4.9 and 4.11) with changing stoichiometry is the square resistance R_S. Henrich demonstrates that R_S decreases with increasing sputter current and therefore, with higher Niobium content. A decreased R_S in Eq.4.11 leads to an increase of the potential barrier U. Therefore, more Niobium content should reduce the dark count rate.

The samples under investigation have the following parameters:

Table 4.3: Parameters of NbN films with varied stoichiometry. I_{sp} is the sputter current in the film deposition process. The Niobium content in an NbN film enlarges with increasing sputter current.

detector	I_{sp}(mA)	d(nm)	$R_S(\Omega)$	j_C(@4.2K)(MA/cm^2)
1	100	4.7	510	3.3
2	145	4.2	430	4.6
3	175	3.9	291	6.0
4	190	4.4	254	4.1

All samples have a nominal width of 100 nm and are embedded in a coplanar waveguide as described in section 4.1.2. The active area is for all samples $4\cdot4$ μm^2.
The experiment is executed in the cryogenic dip stick as described in chapter 3. The operation temperature is 4.2 K. Each of the analyzed samples is measured in a separate cooling cycle. All four samples are measured with closed fiber input. The mounting is realized according to the investigation of the last section with silver paste and almost identical material amount, to minimize a possible influence due to a variation of the thermal interface. Nevertheless some of the counts could origin in black body radiation and are not pure dark counts because the fiber is still mounted during the experiment above the detector. The different stoichiometries lead to slight differences in the critical current I_C even for identical temperature conditions. Therefore, all measurements are plotted vs. I_B/I_C to enable comparability. The measurements were done in part by Dagmar Henrich, Steffen Doerner and the author.

Figure 4.26 shows the dark count rates over I_B/I_C for all tested stoichiometries. Detector 1 is the sample with highest Nitrogen content and detector 4 has the lowest Nitrogen content. Detector 4 has definitely the lowest dark count rate which is 3 orders of magnitude smaller than the highest dark count rate at I_B close to I_C. Also the value of I_B, where DCR overcomes the background varies between the samples. Detector 1 has already DCR of more than $1\ s^{-1}$ from $I_B = 0.75\cdot I_C$. Detector 4 reaches this rate from $I_B = 0.95\cdot I_C$, which is in the range of a typical operational current.

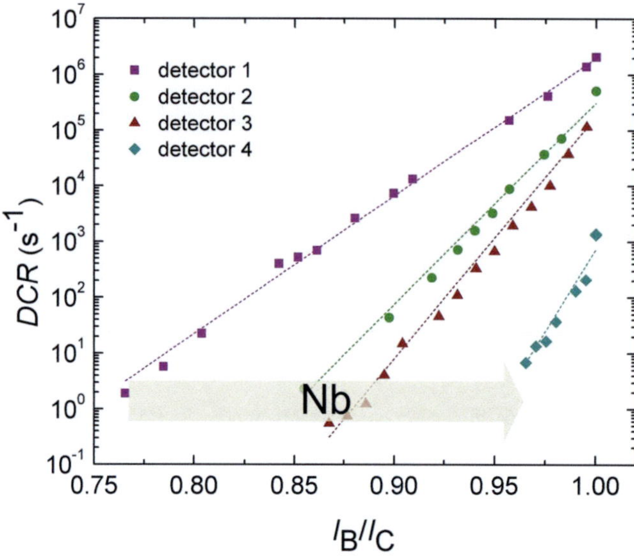

Figure 4.26: Dark count rate for different stoichiometries. [HDH$^+$12]

Based on the vortex model, the change of R_S which is the consequence of the variation of the stoichiometry (see Table 4.3) should directly change the potential barrier. With $U \propto 1/R_s$ from Eq. 4.11, and $P \propto exp(U)$, R_s can be interpreted as factor of the denominator of the rational function in the exponential function in Eq. 4.10. A change of R_S leads to an exponential dependence between the slopes of the curves in Fig. 4.26. The extracted denominators of the exponential fits in Fig. 4.26 are plotted versus R_S in Fig. 4.27. The denominator in the exponent obviously shows a linear dependence on R_S, which is in consistence to the vortex model.

4.2.5 Comparison of all three types of modifications of *DCR* versus influence on detection efficiency

Decreasing the dark count rate is the main focus of the last experiments, but simultaneously the detection efficiency should be the same, or at least not smaller after the process of improving the dark count rate.

For the thermal interface variation experiment, an analysis of the spectral dependence of the system detection efficiency (SDE) at different cooling interfaces at $I_B = 0.95 \cdot I_C$ and $T = 4.2$ K is arranged. The identical 4 nm detector, which was used for the dark count measurement is used again for an *SDE* measurement once mounted with silicon grease and once mounted with conductive silver. The *SDE* is defined by the number of pulse counts divided by the number of incident photons as explained in section 2.2.1. The *SDE*

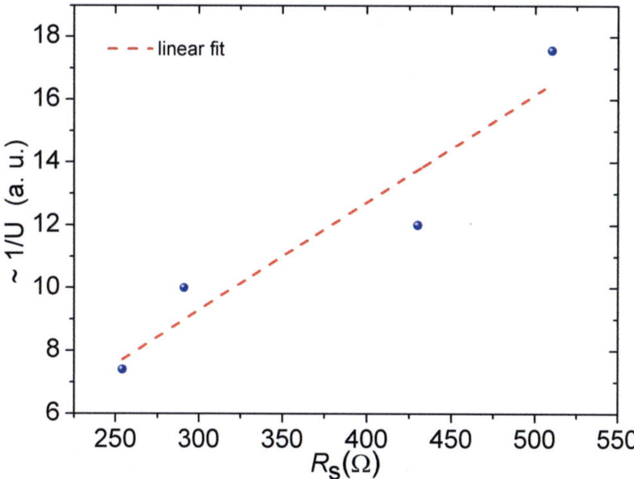

Figure 4.27: Extracted denominator of the exponential fits of *DCR* for different stoichiometries of NbN over R_S from the curves in Fig. 4.26.

measurement allows a qualitative comparison although not all optical losses in the setup are accounted for.

Fig. 4.28 shows the typical shape of the *SDE*. At shorter wavelengths, there is an increase of the *SDE* to the increasing absorptance of the film up to a maximum at around 500 nm. The *SDE* has a continuous decrease in the infrared part of the spectrum. Although there is a systematic difference at larger wavelengths between efficiencies, which are measured for the two different interfaces, it remains relatively small. Moreover, the cutoff wavelength, which is strongly influenced by the temperature T, the detector material and the selected bias current I_B is for both measurements identical. Therefore, to the first approximation, the variation of the cooling strength of the interface does not practically affect the detection efficiency.

A change in the thermal coupling suggests also an influence of the detector temperature T. Assuming, the improved thermal coupling would reduce by a δU the vortex barrier, intentions to decrease *DCR* are contradictorily to the photon detection because a decrease of the vortex barrier, intending a reduction of *DCR*, would also decrease the probability of a infrared detection. The improved thermal coupling should decrease the *SDE* in the infrared range, which is not visible in the measurement (the slight variation in the decay in Fig. 4.28 shows even an improvement for the better cooling.

The investigation of the dark counts versus cooling interfaces already suggest that the temperature is constant in all cases. The difference lies in the type of activation energy. The thermal fluctuations in the film may originate from different points in the system. They are not automatically localized in the electron system of the superconducting meander, and the improved cooling tries to prevent them influencing the vortex barrier. In the photon detec-

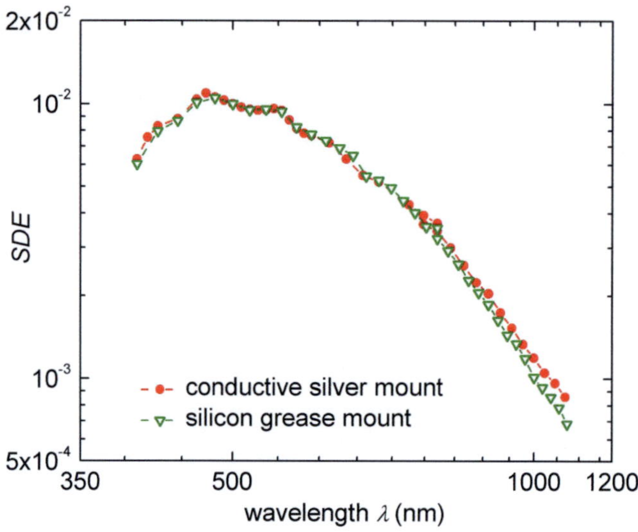

Figure 4.28: System detection efficiency for a 5 nm thick detector mounted with silver paste and
silicon grease. [HRI$^+$12]

tion process, the absorbed photon energy is directly localized in the electron subsystem of
the superconductor and can be considered as local temperature peak in the thin film. The
better thermal coupling of the substrate does not improve the out-banishing of the additional
energy. Therefore, the detection efficiency stays un-affected.

The temperature measurements show a comparable situation. DE improves with lower
temperature. The reason for that is partly a still open question because a reduced operation temperature actually increases the energy gap Δ. Therefore, the cutoff should move
to shorter wavelengths. Measurements show the contrary result. Engel gives one possible
explanation, based on a temperature dependence of the electron diffusivity D [EIS$^+$13].

Concerning the change in stoichiometry, Henrich demonstrated in [HDH$^+$12] that the
spectral width of the detection efficiency improves due to a spectral shift of the energy cutoff
with a larger Niobium content. In Table 4.3 the square resistance R_S decreases with a larger
Niobium content, which leads to a decreased DCR. This should actually lead to a shift of the
energy cutoff in direction to shorter wavelengths ($\lambda_C \propto 1/N_0 \propto 1/R_S$). However, Henrich
showed that the change in stoichiometry towards more Nb also decreases the energy gap Δ
and increases the diffusion coefficient D, which both lead to a contrary effect [HDH$^+$12].
In Table 4.3 one can see that the current density j_C increases for the samples with higher
Nb content. It was already discussed that the measured critical current I_C is smaller than the
expected deparing current I_C^d (see section 4.1.4.2 and [HRH$^+$12]). Therefore, the higher j_C
enables a closer relation I_B/I_C^d, which leads to an improved detection efficiency and a shift
of the cutoff wavelength towards the infrared regime.

However, one has to clarify that the samples with more Niobium content are still far away from a pure Niobium film. Drawing conclusions from these results should not be taken for direct comparison of Nb and NbN detectors. However, one can conclude, the improved detection efficiency for higher content of Niobium coincides with a tremendous decrease in the dark count rate [HDH$^+$12].

One can conclude that all three concepts described in this section enable an improvement of the dark count rate. Moreover, the dark counts seem to follow the vortex model as in the infrared case. The reduction of the dark counts agrees in all three cases either with an equal detection efficiency or even an improvement of the detection efficiency.

4.3 Summary

In this chapter, the intrinsic behavior of SNSPDs is discussed which influences strongly the performance of SNSPDs. The variation of the detector geometry offers a suitable set of parameters to analyze dependencies of performance parameters of SNSPDs. The strong impact of the geometry on superconducting and other parameters, like critical current I_C, energy diffusion length and heat volume affects the intrinsic processes in the detector. In this chapter, mainly the thickness of the superconducting film is under investigation. First an intense study of the pulse response is realized to understand its dependence on spectrum, thickness and bias current. The spectral detection efficiency is investigated with regard to the dependence on bias current and thickness. The obvious decay in the spectral response is discussed in context with the hot-spot model. Finally, the spectral intrinsic detection efficiency is evaluated. It is shown that the intrinsic detection efficiency is 70 % in the maximum, and it is figured out that the absorptance is the most limiting factor of the detection efficiency in SNSPD detectors. Since the hot-spot model is not able to explain the decay of the detection efficiency in the infrared regime, a vortex detection model is considered as suitable explanation for infrared detection and therefore, enforced as suitable detection model. Additionally, the dark count rate is in the focus of this chapter. Several concepts of reduction of the dark count are discusses. The concepts concentrate on an improved thermal interface between the cooling bath and the detector, a variation of the operation temperature and a changed stoichiometry of the superconducting material. With all three methods a tremendous reduced dark count rate could be reached, keeping the detection efficiency constant or reaching even an improvement. All three concepts are compared to the predictions of the vortex model, which is assumed to be the origin of the dark count evolution.

5 Readout concepts for SNSPDs multipixel arrays

Since up to now most of the SNSPD developments are realized by a single-pixel concept, new concepts are required to enable the cryogenic readout of an SNSPD multipixel chip. In this chapter, two concepts of SNSPD multipixel readout are introduced and discussed, which are suitable for imaging applications and real-time single-photon evaluation of an SNSPD array.

5.1 Multipixel systems

The basic aim of imaging is to resolve a multipixel matrix. In principle, a single pixel is able to resolve a picture, by using a scanning technique. The picture, which is projected to the detector plane, can e.g. be scanned by a single pixel by guiding the pixel along a row of the projected picture, which is sequentially repeated row by row until a complete two-dimensional information is recorded. This method has two main disadvantages. First, it can not measure the optical radiation on all pixels in parallel, which means, it measures only the radiation in a short time slot. The second point is, the system requires an exact mechanical construction to enable the scanning functionality.

An improvement of the system would be the use of a multipixel array in combination with a scanning system. Independently, if the array is ordered in a straight line or alternative shape, it allows to reduce the scanning time for one picture. However, a mechanical scanning is still required.

The best solution is a direct measurement with a multipixel detector, whose number of detector elements is equal to the required pixel resolution of the final picture. This allows real-time imaging, continuous observation of each pixel and requires no additional mechanics.

Multipixel systems for single-photon detection, consisting of room temperature detectors like a CCD chip or of an array of avalanche diodes, already exist, but they have certain limitations. The limitation of CCD chips is mainly the long integration time for low photon rates or inversely considered, a high readout noise for fast readout speed. Avalanche diodes on the other side have dead times of up to several tens of nanoseconds, which means that the detectors are inactive after each photon event for a certain time. Generally, the spectral detection bandwidth of the CCD and avalanche detector is defined by the energy gap of the semi-conductor material, which is limited in both cases.

An SNSPD has a broader spectral bandwidth. It can cover with one single structure a spectral bandwidth (400 nm to more than 2000 nm), where more than three CCD chips of different semiconductor types are necessary (see section 4.1.3). Moreover, SNSPDS offer low time jitter t_j [16] and low dark count rates DCR [HRI$^+$12]. These detector parameters are on a competitive level compared to the room temperature detectors. Therefore, the use of SNSPDs in multipixel approaches attracts in spite of the additional cooling costs more and more attention.

An SNSPD multipixel has extended requirements compared to a room temperature system. One crucial point is the operation at cryogenic conditions. To keep the operation temperature at the required level, the number of wires between cryogenic area and room temperature electronics has to be minimized to reduce thermal impact. For the same reason, additional electronic components at the cryogenic stage should have low heat dissipation. Furthermore, the single-pixel performance of each detector pixel should be maintained (high count rate, high detection efficiency over a wide spectral bandwidth, low dark count rate, low jitter) to keep the competitiveness compared to room temperature systems.

The kind of multipixel readout depends on the applications and can be quite different: E.g. imaging in astronomy needs high-resolved pictures with high optical sensitivity but low measurement speed. Applications like spectroscopy require real-time measurements of either a array line or better a matrix of detectors, with high optical sensitivity [57]. In quantum optics [16], arrays of detectors could be helpful to resolve parallel photon events on different channels in real-time. Even multi-photon resolution on one detector element is a quite interesting approach [58].

In this chapter, two concepts are discussed. One concept evaluates the average photon rate of several pixels. This concept is suitable for typical imaging applications and allows to repetitively measure imaging frames with a certain video rate (see Fig. 5.1). It uses the connection of SNSPDs with Rapid Single-Flux Quantum Electronics (RSFQ), which allows digital signal processing at cryogenic temperatures. RSFQ is based on flux quantum data processing with overdamped Josephson junctions. The concept is discussed in connection with multiplexing methods like TDMA, CDMA.

The second concept is about the real-time observation and readout of a detector array by evaluating an individual time-tag signature of each detector element (see Fig. 5.2). This multipixel scheme has a low system complexity at cryogenic level, since the required passive and planar rf structure, which provides the multiplexing functionality, is fabricated directly in the detector production process.

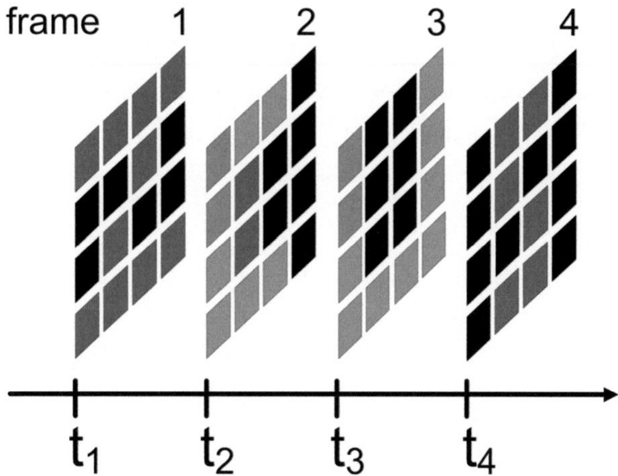

Figure 5.1: Principle of an imaging approach. After a certain framing time the mean count respectively photon rate is evaluated for the full pixel matrix.

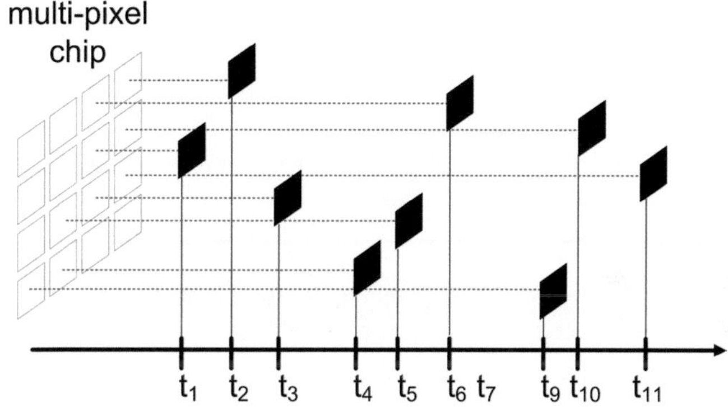

Figure 5.2: Principle of a real-time single-photon readout applied to multipixel arrays/matrices.

5.2 Signal conversion of a single-pixel SNSPD by RSFQ electronics

An SNSPD multipixel concept should work at cryogenic conditions and should have a very low heat entry. In this section, a method is introduced that uses RSFQ to realize digital electronics. This type of superconducting electronics works with a flux quantum as data carrier. RSFQ can be designed for different type of logic functions [115], [116], [117]. The required signal conversion at the input stage of such a logic is described and demonstrated with an SNSPD in this section.

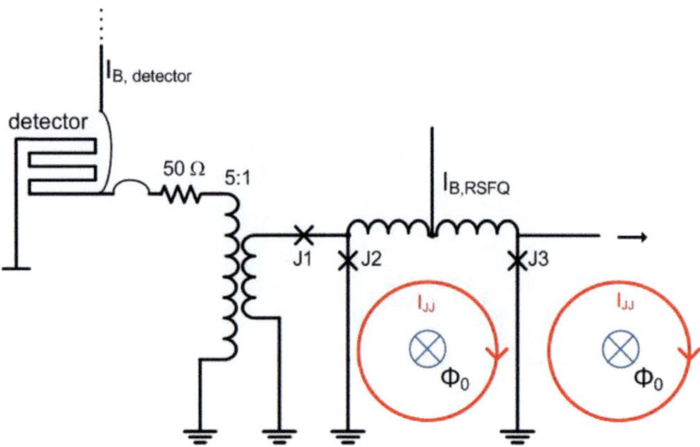

Figure 5.3: Simplified circuit of the DC/SFQ converter.

5.2.1 RSFQ interface for SNSPD readout

The RSFQ logic based on over-damped Josephson junctions (JJ). This electronics allows a processing speed which is faster than required from detector site [118] and offers the common types of logic functionality as in transistor logic. The power consumption of this electronics is quite low, due to the superconducting steady state [119]. RSFQ electronics is a powerful candidate for readout of SNSPDs for the reason that the operation temperature is identical to the SNSPD operation conditions. Moreover, the fabrication processes are compatible. Both facts leads to the possibility that the detector chip and the readout can be located very close in one cryogenic system and even an integrated one-chip or hybrid solution is thinkable in future. The basic aim of an RSFQ–SNSPD readout is a multiplexing in form of a data merging or a data pre–processing at cryogenic level. The data carrier of the RSFQ electronics is a single flux quantum $\Phi_0 = 2.067\,\mathrm{mVps}$ [116]. It is consequently necessary to convert the SNSPD pulse into a single flux quantum (DC–SFQ conversion), which then can be processed in the RSFQ electronics. An SNSPD response can be considered as a digital pulse with random occurrence with an amplitude in the range of 1 mV on a 50 Ω resistance. Each detector pulse redistributes the bias current from the detector path into the readout path due to the high ohmic normalconducting belt in the SNSPD (see Fig. 5.3). This change of current can be interpreted by an RSFQ- DC/SFQ converter unit by triggering a magnetic flux quantum at the output. The concept of such an SNSPD-RSFQ interface was first proposed and also realized by [59] and realized by [OHF+11]. The system from Ortlepp is used in this thesis.

A simplified circuit overview of this proposal and its functionality is given in Fig. 5.3. A more detailed description of this circuit can be found in [OHF+11].

The SNSPD is connected to the input of the RSFQ electronics and biased with a current $I_{B,RSFQ}$. In steady state a external applied bias current $I_{B,RSFQ}$ is split and flows partly

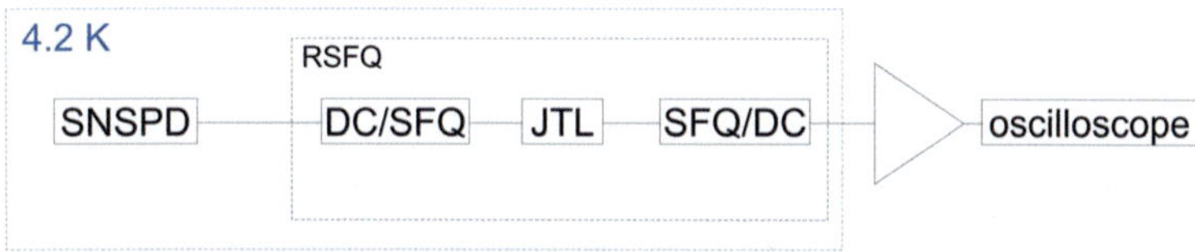

Figure 5.4: Scheme of the signal chain for the SNSPD readout of a single pixel by RSFQ electronics. JTL is a Josephson transmission line.

through the junction $J2$ reaching around 80% of the critical current $I_{C,RSFQ}$ of $J2$ and a small current is flowing through $J1$. The splitting is defined by the inductive divider of the second transformer coil and the inductance of $J2$. A photon event of the SNSPD redistributes the bias current $I_{B,det}$ in the input coil of the transformer stage of the RSFQ circuit. This current is multiplied by the transfer ratio of the transformer stage. This injected current flow is in opposite direction to the already existing current flow through $J1$, which leads to a reduction of the superposed current through $J1$. The injected current flows through $J2$ to GND and adds to the already existing current through $J2$. The current through $J2$ exceeds the critical current I_C. This current trigger releases a single-flux quantum in the junction $J3$, which leads to a loop current through $J3$ (see red circles in Fig. 5.3) and the following RSFQ logic unit. The single-flux quantum can be further processed in the logic unit, which follows the DC/SFQ stage (not depicted in the scheme). Moreover, due to the magnetic induction, the current trough $J1$ increases, as well. If the detector pulse decreases again, the current through $J1$ increases until $J1$ switches, consequently the circuit returns in the steady state and is ready for an new pulse event. $J1$ is required to protect the SNSPD from current reflections of the RSFQ electronics as a back lock.

The RSFQ interface is designed based on the requirements of an SNSPD response. The design used in the experiment has a $5:1$ transformer at the input of the RSFQ electronics and a bandwidth of 8 GHz. The RSFQ electronics requires a minimum current of 25 µA at the input to trigger a flux quantum. A higher current would even provide lower error rates at the output. Compared to typical I_C values, this is not low enough for SNSPDs with high detection efficiencies [HRI$^+$10b], but sufficient for a first feasibility test.

To test this concept in a real measurement, the RSFQ input stage requires a SFQ/DC conversion to record the data after amplification with a standard room temperature electronics like an oscilloscope. The output function of the RSFQ electronics consists of a toggle flip-flop, which toggles its state with each incoming pulse event. To get a voltage signal again, a SQUID device follows. The details of the SFQ/DC are not described here, but can be found in [120].

The fabrication of the RSFQ chip is done by using the standard Fluxonics foundry process for superconducting electronics at IPHT Jena, Germany. The typical fabrication parameters of this process can be found in [121]. The complete signal chain is depicted in Fig.5.4. One

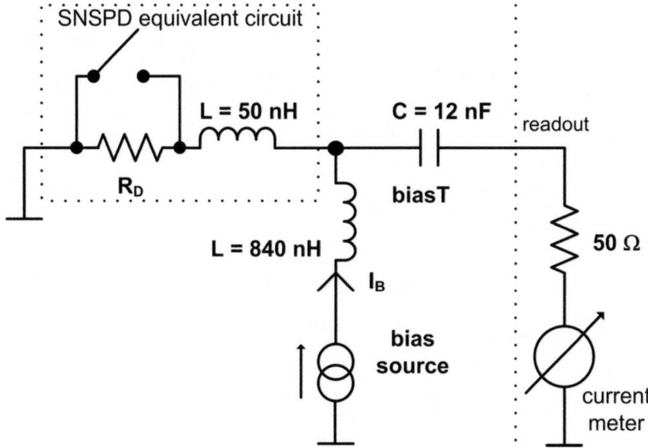

Figure 5.5: Simple electrical model of the SNSPD with connected readout.

has to estimate, how much current of the applied detector bias current I_B is redistributed from the detector path to the RSFQ input to select a suitable detector geometry. The SNSPD and the readout path form a parallel circuit. In the moment of a pulse event, the bias current is distributed in both paths. The exact distribution depends on the maximum impedance of the normalconducting belt in the detector meander in relation to the 50 Ω input impedance of the RSFQ circuit.

The distribution can be approximated with a simulation of an SNSPD response. The simulation is realized by Advanced Design Software from Agilent. The detector is modeled by a switching resistance R_D in series with a inductance $L = 50$ nH (Fig 5.5). The pure switching of R endures 200 ps, which is in the range of the detector relaxation time τ_{es} [RPH$^+$10]. Both current paths are separated by a 12 nF capacitor C. The bias current is set to $I_B = 42$ µA, which fits to a 5 nm thick film with nominal meander width of 100 nm [HRI$^+$10b]. The readout path consists of a 50 Ω impedance. The maximum resistance R_D is varied between 100 Ω, 500 Ω and 2000 Ω. In Fig. 5.6 both the simulated current in the detector path and the simulated current in the readout path are shown. One can see that the detector belt resistance R_D requires a value of more than 2000 Ω to redistribute the current completely in the readout path. In real detectors the belt resistance varies due to the meander geometry, film parameters like square resistance R_S, the film thickness d and linewidth w. The hot-spot resistance at the beginning of the pulse event can be assumed to be in the range of 50 Ω for a 4 nm thick film, however Joule heating tremendously growths the normalconducting belt during a pulse event [20]. That means, it cannot be compulsorily assumed that the complete bias current I_B is redistributed into the readout path.

It should be noted that all the design parameters of the SNSPD influence the performance parameters like detection efficiency and the dark count rate, as well (see chapter 4). Therefore, one has to select carefully the appropriate detector parameters to reach both, a high

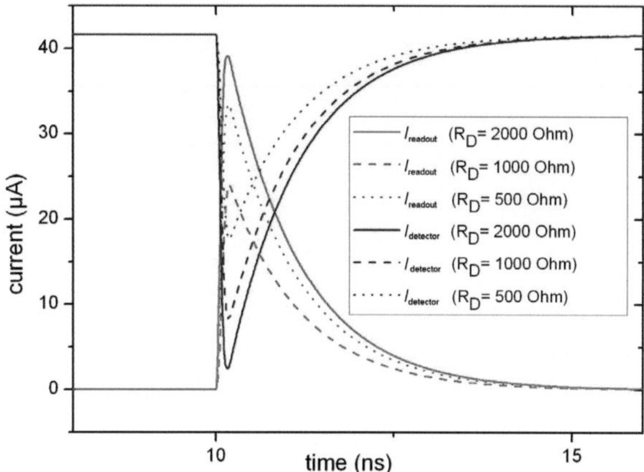

Figure 5.6: Simulated time domain response of the currents in the detector circuit for different normalconducting belt resistances R_D.

current distribution from sufficient I_B to trigger the input of the RSFQ, and a high detection efficiency, which normally operates at low I_B.

5.2.2 "Proof of principle" of an RSFQ input stage for SNSPDs

The following results base on the work published in [OHF$^+$11].

To avoid a limitation of operation due to a too low current at the input stage of the RSFQ electronics, a detector with thickness $d = 10$ nm is fabricated, which has a critical current I_c of 136 µA at 4.2 K. The SNSPD meander has a width $w = 100$ nm and is provided with the standard coplanar readout as used in the chapters before. The SNSPD and the RSFQ chips are mounted on a common brass block close to each other (see Fig. 5.7). The SNSPD rf output is directly connected to the pads of the RSFQ input by aluminum bonds. The detector bias line is equipped with a 850 nH coil as part of a biasT with separate wires to room temperature. The RSFQ electronics needs five biasing wires for operation to control each stage of the circuit, separately. Two of them allow to set the input stage. The other lines control the merger function, which is used later in the experiment, and the output toggle flip-flop. The chips are fixed on a rf circuit carrier board made from Rogers TMM10I substrate. The board contains the dc lines and the rf-output line, which is connected to a PCB-SMA connector. Both chips have separated GND lines to prevent a dc cross-talk by potential shifts along the wires. They are connected to individual current sources, which provide the required currents. Only non-magnetic components are used for the cryogenic setup and magnetically shielded by a Cryoperm tube.

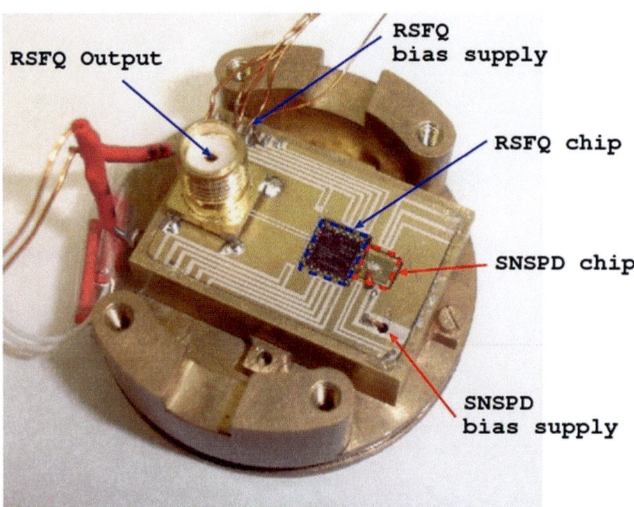

Figure 5.7: Mounted SNSPD and RSFQ chip on a brass holder. The chips are enclosed by an rf board for rf signal guiding and dc biasing.[OHF$^+$11]

In a first experiment, the operation point of the RSFQ electronics is defined, which is set by $I_{B,RSFQ}$ of the first stage of the RSFQ circuit. The SNSPD is biased at $I_{B,det}$ of around $0.75 \cdot I_C$. In this mode no dark counts appear. A 650 nm laser source is modulated with moderate frequency by a rectangular function to get defined light "on" and "off" intervals. The optimal operation point of the RSFQ circuit is found, if the number of togglings during " light on" reach a maximum and no togglings appear in the " light off" phase (a too strong $I_{B,RSFQ}$ of the input stage of the RSFQ electronics can set the sensitivity to an operation point, where error togglings are produced). Figure 5.8 shows the situation of optimal operation.

During the modulation of the light, the complete circuit was operated only at one fixed bias current of the detector. Since the amplitude of the current pulse at the input of the RSFQ transformer depends on $I_{B,det}$, an analysis of the counts dependent on $I_{B,det}$ is required. Figure 5.9 shows that the counts extracted from the RSFQ output signal have an exponential dependence on $I_{B,det}$. This fits to the typical behavior of an SNSPD between 60% and 90 % of I_C. The typical plateau of the counts close to I_C does not exist for an 10 nm thick detector at temperatures around 4.2 K [HRI$^+$10a]. The stop of the pulse conversion due to the limited input sensitivity of the RSFQ circuit can't be investigated in this measurement because the $I_{B,det}$-dependent detection efficiency is already to low, to provide pulse events at this low ratio $I_{B,det}/I_C$.

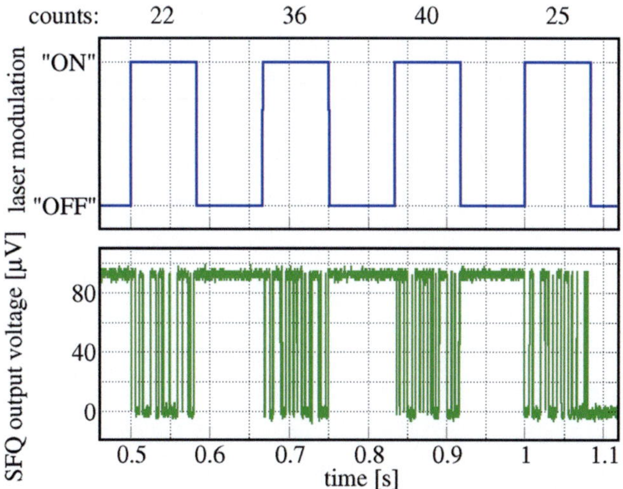

Figure 5.8: Top panel: Laser state defined by a rectangular function. Bottom panel: The RSFQ output signal depending on laser state.[OHF$^+$11]

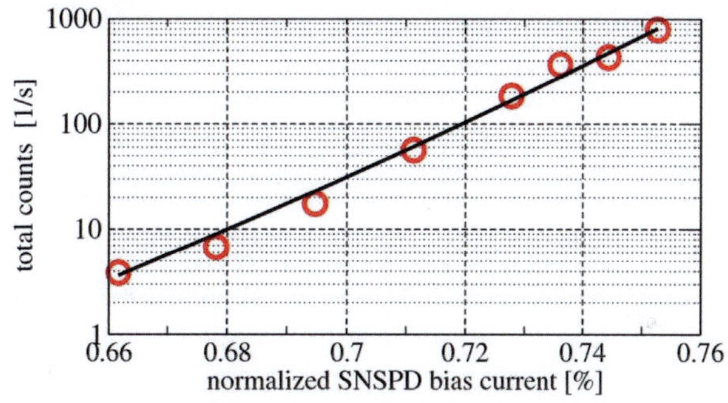

Figure 5.9: Count rate over bias current $I_{B,detector}$ measured with the RSFQ electronics.[OHF$^+$11]

Alternative SNSPD-RSFQ readout with oversampled synchronous data readout

The RSFQ signal conversion described in the last section mainly suffers by the limited sensitivity. Therefore, Ortlepp [OHT$^+$13] introduced an innovative SFQ conversion concept, which allows an input sensitivity of 2 µA. The concept was sucessfully tested by Ortlepp and the author. It is not in the focus of this work to describe this circuit and the experiment in detail. However, a short overview demonstrates further directions of an SNSPD/RSFQ readout.

The concept is based on the idea to digitize the analog pulse of the SNSPD response by an oversampled conversion with a synchronous applied rectangular function, like in a delta-sigma converter [122]. The core functionality is realized by an RSFQ-based comparator function. The JJ in the comparator acts as logic unit and at the same time it operates as current generator for the SNSPD. An external bias source is not necessary any more. The detector bias can be controlled by the frequency of the rectangular function, which is externally applied to the RSFQ circuit.

This new input conversion stage requires the SNSPD on top of the RSFQ circuit (see Fig. 5.10) to keep the wire connections as short as possible, which reduces the impedance of the wire bond.

The readout operates in the following way: The RSFQ comparator is supplied by a dc $I_{B,\,RSFQ}$ which leads to an operation in the grey-zone of the comparator [123]. The additionally applied rectangular frequency provides only partly a toggling of the comparator output. A normalconducting belt leads to a redistribution of the bias current from the detector path into the comparator stage. The operation point of the comparator is moved out of the grey-zone and the probability of the toggling of the comparator is changed, which leads to a change in density of output toggling depending on the superposed current strength. The RSFQ signal is converted into a voltage signal by a Suzuki stack SFQ/DC converter, which allows the direct data acquisition without additional bandwidth-limited amplifiers [124]. The first measurements are demonstrated in Fig. 5.11. The red curves indicates the moment of a photon pulse which is applied by a laser diode ($\lambda = 650$ nm). The green curve is the bit chain, which is produced by the RSFQ chip. The bit chain is synchronized by an external clock. In the moment of the optical pulse excitation, the SNSPD produces a response pulse, which changes the density of data bit toggling like in a pulse width modulation (PWM). The resistance in the readout is intentionally selected in a way that the decay of the detector is in the µs range, to reach adequate relation between the sampling rate and the response pulse length. The output signal can be interpreted as an oversampled digitization of the SNSPD

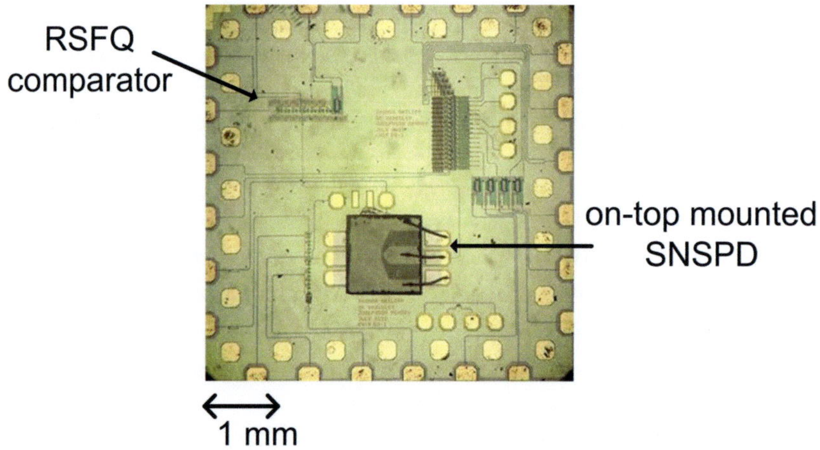

Figure 5.10: SNSPD pixel with $1 \cdot 1$ mm^2 size on top of the $5 \cdot 5$ mm^2 RSFQ readout circuit. The SNSPD is directly wire-bonded to the RSFQ circuit.

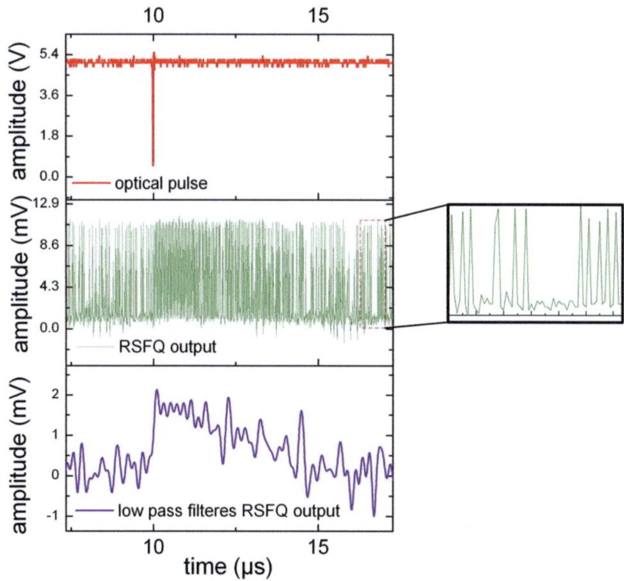

Figure 5.11: Measurement with the new comparator-based RSFQ concept. The red curves indicate the moment of a photon pulse ($\lambda = 650$ nm) that hits the SNSPD. The green curve is the synchronous bit chain, which is produced by the RSFQ chip. At the moment of the optical pulse excitation, the SNSPD produces a response pulse which changes the density of data bit toggling like in a pulse width modulation (PWM). The pulse amplitude can be reconstructed after low-pass filtering (blue curve, $f_{3dB} = 5$ MHz).

pulse. The original pulse shape can be reconstructed, if the digital data output is filtered by a low-pass filter (blue curve, $f_{3dB} = 5$ MHz).

5.3 Multipixel readout with RSFQ electronics for imaging applications

The signal conversion discussed in section 5.2 is the basic requirement to enable data processing of SNSPD signals with RSFQ electronics. A suitable multiplexing logic has to follow to readout several channels and to reduce the number of room temperature lines. The aim of this readout concept is to measure average count rates for high-pixel imaging applications. The maximum video frame rate should be real-time, which means 25 Hz. The following experiments and descriptions base on the work published in [HWE+12].

5.3.1 RSFQ based signal conversion with pulse merger functionality

For a multipixel concept, the already described RSFQ data converter has to be extended by a merger function or an alternative multiplexing functionality. Terai proposed two solutions [59]: One considers the use of a cryogenic shift register, with parallel register inputs for each detector channel. A pulse event of a detector pixel toggles the dedicated register bit to "one". After a certain time the register data will be shifted out by an external clock and guided to an acquisition electronics at room temperature. This method allows the measurement of imaging frames, however, with dead times during the data shifting. Another concept asynchronously merges all RSFQ inputs to one common readout line and puts all SNSPD events sequentially on one output line.

Normally, SNSPDs are biased continuously. Applying this method to a multipixel merger, the readout counts all pixel events but loses the information about the local origin of the count events, for which reason intelligent extensions are required to make such a system suitable for imaging applications.

The readout of a merger system is less complex concerning wiring and operation. Therefore, this variant of the proposals in [59]is realized in these experiments.

A merger function, also called confluence buffer (CB), combines two input lines to one common output line, see Fig. 5.12 a). The design of such a merger is built with a similar schematic as in the DC/SFQ converter, however, with two input branches connected in front of $J3$ (see Fig. 5.3) [116], [117]). Using several merger stages and Josephson transmission lines (JTL) , all inputs can be combined to one readout. To minimize the number of mergers in the circuit, a binary-tree structure is suitable as shown in Fig. 5.12 (a). This design minimizes the required number of JJs. In [HWE$^+$12], Ortlepp adds a suggestion for a sequenced order using passive transmission lines (PTL) in between (see Fig. 5.12 b)). Although more mergers functions are necessary in this scheme, it would allow a better layout on the chip plane and consequently would lead to a homogenous pixel matrix.

For the measurements, a 4-pixel merger in form of a binary tree structure is realized. The input stage is realized as described in section 5.2, therefore, the RSFQ circuit has the same input sensitivity.

Terai et al. showed that such an RSFQ merger has the same detection efficiency DE and the same time resolution t_j as classical SNSPD readout concepts [60], [44].

5.3.2 4-pixel detector design

All pixels should be closely located on one chip to increase the filling-factor of the active chip area and to enlarge the optical coupling to the detector chip. Therefore, a pixel matrix on one chip is quite reasonable, which allows a pixel mapping in a row and column system. The distance between the pixels depends on the required image resolution defined by the diffraction characteristics of the system. If the image light is guided by a bundle of fibers into the cryostat, the minimum distance between the pixels is defined by the maximum resolution of the fiber bundle $\Delta x = 2d_{core}$. d_{core} is the fiber core diameter. If free-space coupling is used, the minimum distance is defined by $\Delta x = l \cdot w_{min} \approx l \cdot 1.22\frac{\lambda}{D}$. l is the distance between detector and imaging lens, λ is the highest wavelength that will be measured and D is the

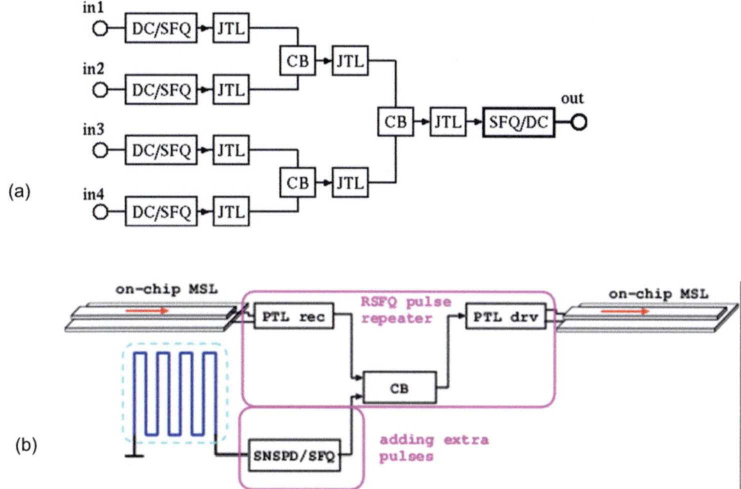

Figure 5.12: Scheme of two variants of SFQ-signal merging. a) Merging of signal channels by a binary tree structure. CB is the confluence buffer (merger) and JTL the junction transmission line. b) Merging each pixel between on-chip passive transmission line (PTL) converters allows the flexible use of micro stripe lines (MSL), which enables flexible routing and placing of the detectors. [59], [HWE$^+$12]

diameter of the imaging lens [36].

A row of detector elements can be fabricated with quite narrow pixel distance [44], and used as base element of a matrix scheme.

For this experiment, a 4-pixel chip, fabricated at IMS, is used to demonstrate the feasibility of a multipixel readout. The NbN film thickness is $d = 10$ nm and the meander width is $w = 100$ nm. The typical detection efficiency of such a detector is quite low [HRI$^+$10b] but for this experiment still sufficient.

The detector elements are located on the chip in a matrix scheme with a distance of 500 µm in between (Fig. 5.13). This distance is not optimized for high filling-factors or for highest imaging resolution, since the light is guided by only one multimode fiber.

The four pixels are completely independently routed on the chip and embedded in four individual coplanar waveguides. The coplanar readout line has the parameters as described in section 4.1.2. All coplanar waveguides are routed to one side of the chip.

5.3.3 Cryogenic setup for time-gated SNSPD-RSFQ measurements

For this experiment the cryostat system from chapter 3 is used. The two chips are mounted on a modified chip holder, which allows placing the SNSPD and the RSFQ chip close to each other (see Fig. 5.14). The SNSPD is mounted by conductive silver paint to achieve a high thermal coupling. The RSFQ chip is mounted by thermal varnish, which prevents electrical short cuts and guarantees nevertheless a high thermal coupling.

The SNSPD is positioned below the fiber end of the multimode fiber and is illuminated by a $\lambda = 650$ nm free-space laser diode. The light spot on the detector chip represents the photon distribution of the interference of the laser modes, which leads to different photon flux on the pixels.

In the multipixel experiment the cryogenic setup is extended by an additional room temperature part. An overview is given by Fig. 5.15. The room temperature electronics is divided in two parts: a data acquisition part and a current switching part. The data acquisition part was prepared by the common work of Olaf Wetzstein from IPHT Jena and Benjamin Berg from IMS, Karlsruhe.

Data acquisition with the FPGA board

The data acquisition system is an FPGA of the type Cyclone 3 with an internal 50 MHz clock. The FPGA input port is localized directly behind a room-temperature differential amplifier (bandwidth: DC to 250 kHz, 40 dB gain). The data input is converted by an 18 bit analog/digital converter (ADC). This ADC allows a maximum resolvable count rate of 200,000 counts per second. For future, this rate can be increased by using an FPGA board with higher clock frequency or using the digital port of the FPGA as direct input without ADC in between, which enables a clock rate of 10 MHz. The FPGA contains an edge-detection function, which registers in the digitized data stream the togglings of the RSFQ output. In the current configuration, the FPGA is able to resolve a minimum peak-to-peak amplitude level of 150 µV. The current FPGA firmware is able to count the events in a given time interval t_{clk} by a $32-$ bit counter. The free-selectable parameters for the user are the discriminator level of the edge detection, the sampling rate of the ADC and the time interval t_{clk} of one measurement frame. The FPGA is able to repeat the measurement to enable real–time video frame rates. The FPGA is connected to a computer by Ethernet and can be controlled by a graphical user interface developed at IPHT Jena, which shows the measured real–time data, as well.

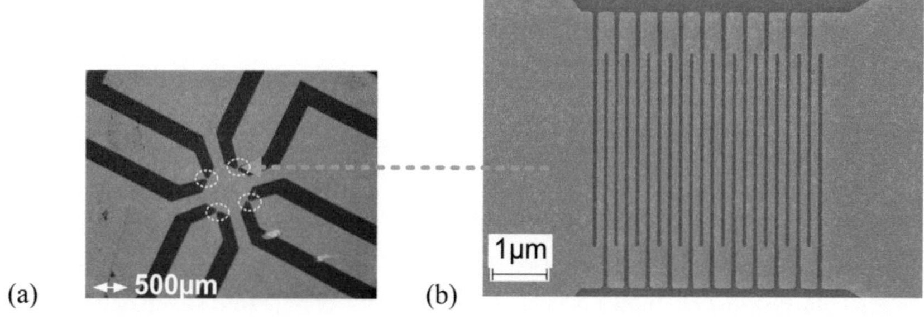

Figure 5.13: Design of a 4-pixel SNSPD made from a 10 nm thick NbN film. a) Image of 4 detector elements with individual coplanar lines. b) SEM image of one single detector element. [HWE$^+$12]

Microcontroller controlled bias switching

Since all channels are merged to one common output line, it is necessary to realize a time-gated measurement by a separate switching of each detector, which allows to set a detector "on" or "off". The easiest way is to switch the bias current $I_{B,\,det}$ of each detector.

In the setup, room temperature CMOS switches are used. The switches are controlled by a microcontroller. Each current switch can be programmed with an individual sequence of "on" and "off" intervals, the so-called states. The states are clocked to the switches with a common reference clock t_{clk} from the FPGA, which is necessary to have simultaneous changes of the states. It is necessary to use switches with low charge injection. These components prevent the detectors from unwanted latching due to voltage transients on the bias path.

Each detector is biased by a separate current source. In operation, the bias source continuously applies a bias current at a fixed value. Depending on the switching state, the bias current flows through the detector or is shorted to GND and is addionalally interrupted in the detector path.

The room-temperature switching may be limited for larger pixel numbers because each detector element needs its own wiring. Even with twisted pair wires of woven loom, the thermal impact increases for high pixel numbers. Brandel et al. proposed superconducting switches which can be used for a cryogenic switching in future [125]. With these switches the number of wires for detector biasing reduces to one single voltage biasing line.

5.3.4 Limitation of the counting accuracy of an SNSPD-RSFQ measurement

Measuring the photon number in a gated time interval, the photon distribution has an important impact on the accuracy of a count rate measurement. It is known that a process with

Figure 5.14: Multipixel SNSPD and RSFQ chips embedded in a carrier board. SNSPD/RSFQ connections and supply lines are realized by wire bonding. [HWE$^+$12]

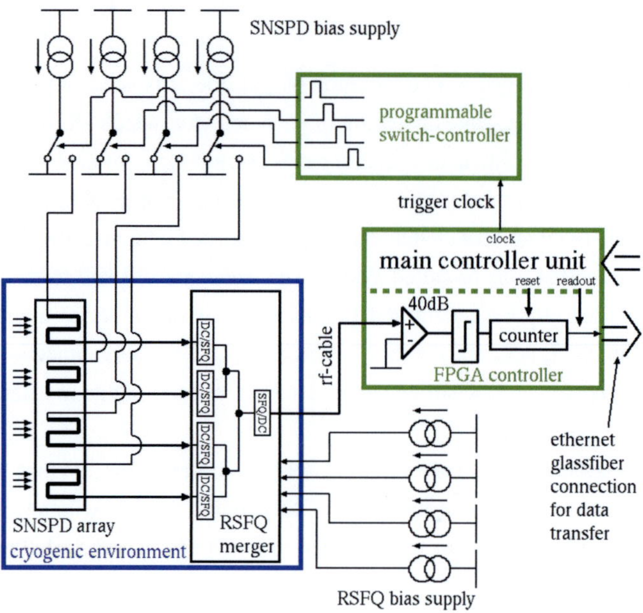

Figure 5.15: Setup for SNSPD measurements with RSFQ readout. The core of the setup (blue box) is the cryostat with the SNSPD and the RSFQ electronics. The code sequences for the bias switching are controlled by a micro controller (upper green box). The FPGA (lower green box) counts the incoming toggling sequences from the RSFQ electronics and controls the synchronization of the measurement. [HWE$^+$12]

single events in a certain time interval follows a Poisson distribution [19]. The probability distribution is given by:

$$f(k,\lambda) = \frac{\lambda^k \cdot e^{-\lambda}}{k!} \tag{5.1}$$

Since a typical SNSPD is not able to resolve multi-photon events, the following analysis is focused on a measurement done in the single-photon regime. The single-photon regime is defined in Eq. 5.1 by $k = 1$. The expectancy value of the Poisson distribution is λ which is the mean value of counts during a measurement time interval. In the Poisson distribution, the standard deviation is $\sigma = \sqrt{\lambda}$.

The mean value λ of counts is given in an SNSPD context as:

$$\lambda = DE \cdot n_{ph} \cdot \Delta t. \tag{5.2}$$

DE is the detection efficiency of a detector element and n_{ph} is the photon flux on the detector element, which is constant in this first simplified consideration. Δt is the measurement interval. Eq. 5.2 shows that the measurement time Δt in connection with $\sigma = \sqrt{\lambda}$ will have a strong influence on the accuracy of a time-gated SNSPD measurement.

For imaging, it is important that the detector readout does not increase the dispersion of

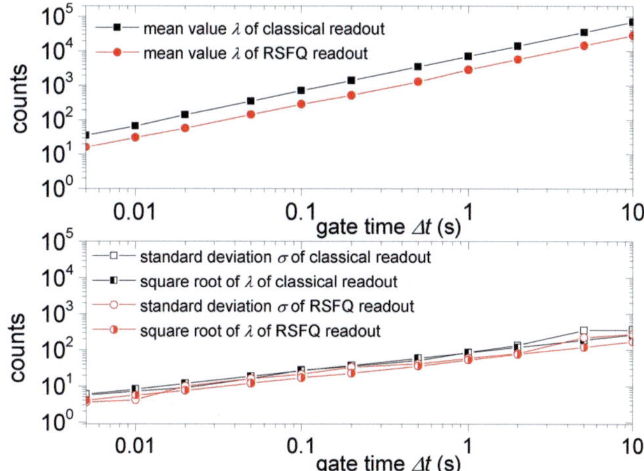

Figure 5.16: Comparison of the count statistics between a classical readout and the RSFQ readout. The standard deviation follows in both cases the square root of the mean value as predicted from the Poisson distribution. [HWE+12]

a count measurement by additional noise counts or other influences. Ideally, the dispersion should be purely defined by the photon distribution and not be increased by the readout electronics.

We check this in an experiment. The counts for different gate times Δt are compared, between a classical pulse counter readout and the RSFQ readout.

First, we measure with the typical pulse counter readout using the SR620. A detector is used with a standard meander width of $w = 100$ nm and a thickness of $d = 4$ nm. The detector is illuminated by light with 400 nm wavelength. The optical power on the detector is ≈ 34 aW (re-calculated from the known detection efficiency and the measured count number). The detector is biased with $I_B = 0.9 \cdot I_C$, which leads to a detection efficiency of 10%.

Fig. 5.16 shows the number of counts (black squares) over gating time Δt as mean value of 30 measured samples. The counts linearly increase with Δt. This is expected from Eq. 5.2. The standard deviation σ is calculated, as well, and depicted as open squares. The measured standard deviation σ is identical with the square root of the mean value $\sqrt{\lambda}$ (half open black squares), which means that, using a standard counter electronics, the data accuracy of the photon counting experiment is defined by the Poisson distribution of the photons.

The experiment is now repeated with the RSFQ readout. To compensate the lower detection efficiency of the multipixel chip, the photon flux is increased by increasing the laser power to operate in a comparable count range. One detector of the multipixel chip is biased with $I_B = 0.9 \cdot I_C$, as well, and the gate times are identically selected as in the experiment before. The red filled circles present now the mean value, which again linearly increases with gating time Δt. Moreover, the standard deviation σ and the square root of the mean value $\sqrt{\lambda}$ are identical. Consequently, the RSFQ readout represents the same characteristic

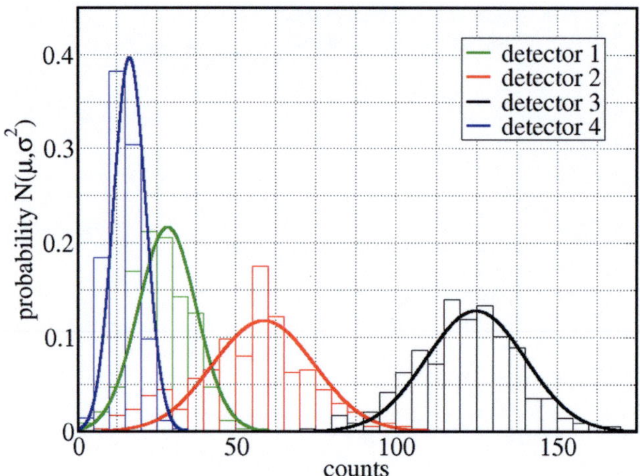

Figure 5.17: Count distribution of all four pixels. The measurements are done with a gate time Δt of 100 ms. The number of counts are Gaussian distributed.

as the classical readout.

One can conclude that the Poisson distribution describes the count distribution of the RSFQ readout for all detection efficiencies, all sizes of photon flux and all measurement times. Therefore, in a time gated measurement a quality parameter q_{count} can be defined by the ratio between the mean value λ and the standard deviation σ:

$$q_{count} = \frac{\lambda}{\sigma} = \frac{\lambda}{\sqrt{\lambda}} = \sqrt{\lambda} \tag{5.3}$$

To make q_{count} independent of external factors like photon flux, these λ and σ should be normalized to the photon flux n_{ph} and $\sqrt{n_{ph}}$, respectively.

$$q_{count,norm} = \frac{\lambda \sqrt{n_{ph}}}{\sigma \cdot n_{ph}} = \frac{\lambda}{\sqrt{\lambda}} \cdot \frac{\sqrt{n_{ph}}}{n_{ph}} = \sqrt{\lambda} \cdot \frac{1}{\sqrt{n_{ph}}} = \sqrt{DE \cdot \Delta t} \tag{5.4}$$

A sequential measurement of the counts of all four detector elements leads to the picture of Fig. 5.17. The counts of the four pixels are Gaussian-distributed, as a consequence that the Poisson distribution fades to a Gaussian distribution for increasing values. With the individual check of each input channel in the previous chapter, the RSFQ readout is prepared for a multipixel measurement.

Dependence on the number of pixels

A simple way to check, if all detector elements are working, is to switch all bias supplies "on" and then to switch them "off" one after another during illumination. The extracted

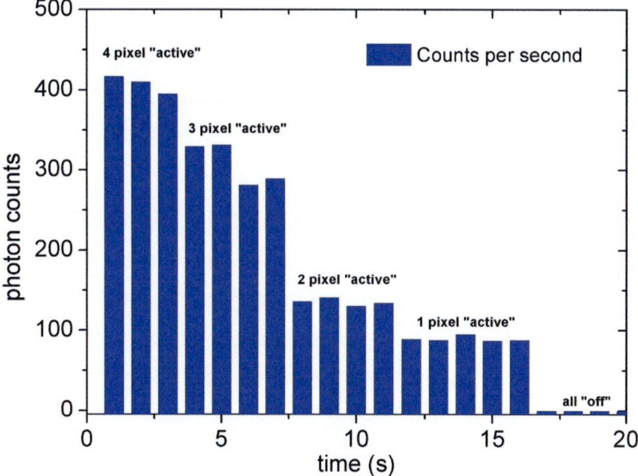

Figure 5.18: Analysis of the measured counts with different number of activated detectors. At the beginning, all detectors are on. After a certain time the detectors are switched off one after another.

counts are shown in Fig. 5.18. The steps of the number of counts after each detector switching are obvious. The total number of counts decreases with each deactivated detector.

The previous discussion demonstrated that a single pixel measurement delivers a certain dispersion, if the measurement time is gated. This is now analyzed for all four pixels. The complete chip is again illuminated by arbitrary photon flux. The number of overall count

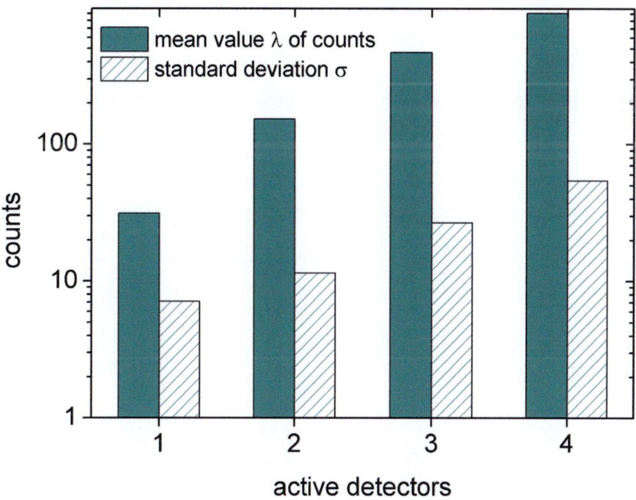

Figure 5.19: Mean value of overall counts and standard deviation for different numbers of active detectors. The standard deviation of the overall counts increases with increasing number of detectors.

events is plotted in Fig. 5.19 for $\Delta t = 50$ ms. The mean value of counts increases with the number of active detectors as well as the standard deviation.

5.3.5 Principle of code division multiplexing access (CDMA) for SNSPD readout

The continuous biasing of all detector elements leads to a loss in the local resolution of the counts because the photon event can not be assigned to a detector element. The easiest way to define the origin is using a time-domain multiple access (TDMA). In this mode the detectors are sequentially biased for a time t_{clk}, and the number of counts are measured and the count rate is estimated. For a given frame rate, the time slot t_{clk} reduces with increasing pixel number. From section 5.3.4 it is clear that the accuracy of a count measurement is worse with shorter measurement time t_{clk}. It would be better to have a concept in which the measurement time t_{clk} is independent of the pixel number.

A possible candidate is a derived version of a code division multiple access (CDMA) approach. Irvin et al. demonstrated for a SQUID readout that such concepts are suitable for detector readouts [126]. The difference to the SNSPD approach is that a SQUID CDMA method multiplexes the analog amplitude of the output, which is assumed to be continuous over time. The CDMA based SQUID readout profits from a reduction of the uncorrelated noise floor, which is typical for CDMA systems [127]. The SNSPD approach requires the multiplexing of random digital events without considering uncorrelated intrinsic noise sources, but with the problem of the dispersion of the count measurement.

The CDMA concept uses the possibility of individual switching of the detectors. In contrary to the TDMA approach, in which all detectors are biased sequentially, the bias lines can also be switched in special sequences. The switching sequences of a detector line follows an orthogonal function compared to the switching sequences of each other lines. Suitable orthogonal code sequences are the Walsh functions, which are given by the Hadamard matrix [127]. The functions can be extracted iteratively from the basic Hadamard $\mathbf{H_2}$ matrix:

$$\mathbf{H_{2n}} = \begin{bmatrix} \mathbf{H_n} & \mathbf{H_n} \\ \mathbf{H_n} & -\mathbf{H_n} \end{bmatrix} \in \{1,-1\}^{2n \times 2n}$$

$$\text{with the starting matrix:} \mathbf{H_2} = \begin{pmatrix} 1 & 1 \\ 1 & -1 \end{pmatrix} \tag{5.5}$$

Only the rows $\mathbf{H_{2n}}(i)$ with $i > 1$ are suitable as sequence for the CDMA approach. Due to the iterative derivation, the order of the Hadamard matrix is defined by $2n$. It means that only matrices with order of power of two are possible. The parameter $2n$ is the number of clocks t_{clk} in one measurement frame. The maximum number of detectors that can be read out is $m = 2n - 1$ because of the canceled row $\mathbf{H_{2n}}(1)$. For higher orders one can define $m \approx 2n$.

Consequently, for a 4-pixel experiment, the Hadamard matrix with order $2n = 8$ is necessary. Table 5.1 shows the dedicated Walsh functions:

Table 5.1: Walsh functions for 4 detectors. The rows represent the clocked switching sequence for each detector. The columns represent the parallel active detectors during one clock interval.

Clock cycle index	1	2	3	4	5	6	7	8
$H_{8,Detector1}$	1	1	1	1	-1	-1	-1	-1
$H_{8,Detector2}$	1	1	-1	-1	-1	-1	1	1
$H_{8,Detector3}$	1	1	-1	-1	1	1	-1	-1
$H_{8,Detector4}$	1	-1	-1	1	1	-1	-1	1
H_{5-8}	...	...	...	...	...	...	...	...

The Walsh states are defined with 1 and -1. 1 means the detector is switched "on" and -1 the detector is switched "off". The rows define the individual sequences for each detector, which is synchronously clocked with the time t_{clk}. Each column represents one t_{clk} interval and determines which detector is "on" in this time.

The multiplexing concept can be described as a matrix/vector multiplication:

$$\mathbf{c_{ext}^T} = \mathbf{c^T} \cdot \mathbf{M} \cdot \mathbf{H_{2n}} \tag{5.6}$$

where the vector $\mathbf{c_{ext}}$ contains in each element $c_{ext,X}$ the extracted counts of the detector element $X \in \{1,2,3,4,...,2n-1\}$ during one acquisition frame $T_{frame} = 2n \cdot t_{clk}$, which is proportional to the photon flux. The matrix $\mathbf{M}$ describes the coding of the channels which is applied by the switching of the bias currents. $\mathbf{M}$ can be formed from a $\mathbf{H_{2n}}$ matrix by replacing -1 by 0. The $\mathbf{H_{2n}}$ matrix in Eq. 5.6 extracts the estimated counts during T_{frame}. Equation 5.6 can be written as:

$$\mathbf{c_{ext}^T} = \begin{pmatrix} 0 & c_1 & c_2 & c_3 & ... & c_{n-1} \end{pmatrix} \cdot \mathbf{H_{2n}} \{-1 \longrightarrow 0\} \cdot \mathbf{H_{2n}} =$$

$$\begin{pmatrix} 0 & c_1 & c_2 & c_3 & ... & c_{n-1} \end{pmatrix} \cdot \begin{pmatrix} 2n & 0 & \cdots & \cdots & 0 \\ n & n & 0 & 0 & \vdots \\ \vdots & 0 & \ddots & 0 & \vdots \\ \vdots & \vdots & 0 & \ddots & 0 \\ n & 0 & \cdots & 0 & n \end{pmatrix} = \begin{pmatrix} n \cdot (c_1 + c_2 + ... c_3 + c_{n-1}) \\ n \cdot c_1 \\ n \cdot c_2 \\ n \cdot c_3 \\ \vdots \\ n \cdot c_{n-1} \end{pmatrix}^T \tag{5.7}$$

Each element $c_{ext,X\{1,2,3,4,...,2n-1\}}$ is only dependent on the counts c_X with identical index

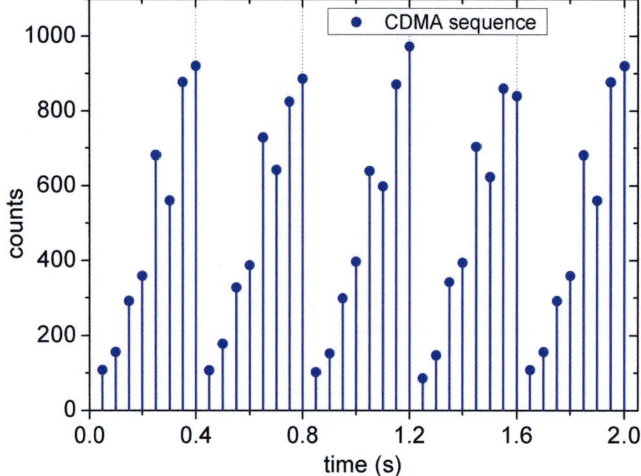

Figure 5.20: Measurement with CDMA switching of the bias currents. Each column represents the number of counts per CDMA state. The CDMA frame is applied five times to the setup.

and therefore, from only one detector element. The original counts are multiplied by the constant prefactor n. The RSFQ readout system delivers a toggling output $Out(t)$, so the matrix scheme has to be modified:

$$c_{t_{clk},X\in\{1,2,...,n-1\}} = \frac{1}{n}\int_0^{T_{frame}} H_{2n,X}(t)\cdot Out(t)dt = \frac{1}{n}\cdot\sum_{i=1}^{2n} H_{2n,X}\left(i\cdot t_{clk}\cdot\sum_{\tau=(i-1)t_{clk}}^{\tau=it_{clk}} Out(\tau)\right)$$

(5.8)

The vector $c_{t_{clk}}$ has in each element the counts collected during the time t_{clk}, which means the n-th part of c_{ext}. The last summation function (right end of the formula) means the counting or collection of the asynchronous digital toggling edges of the RSFQ output module during one clock cycle t_{clk}. $H_{2n,X}$ is the respective row of the detector X of Table 5.1. The mean count rate vector, which contains the final mean count rate information of one imaging frame is then defined:

$$CR_X = \frac{2n\cdot c_{t_{clk},X\in\{1,2,...,2n-1\}}}{t_{clk}}$$

(5.9)

5.3.6 CDMA measurements and analysis of counting accuracy

The orthogonal CDMA sequences of Table 5.1 are programmed in the controller of the bias switching and triggered to the CMOS switches. The switching time t_{clk} can be defined by the control software. The bias current during the active time of the detector is $I_B \approx 0.9\cdot I_C$ for all detectors.

The mathematical description in Eq. 5.7 expects a constant number of counts of each pixel during the frame time T_{frame}. Consequently, measuring several frames one would expect an identical, repetitive pattern of count numbers.

Fig. 5.20 depicts the typical count pattern of an orthogonal bias switching after collecting the counts per time slots t_{clk}. The switching time is selected to $t_{clk} = 50$ ms. The expected eight states of the Walsh code and the repetitive pattern with each frame can be clearly recognized. However, the number of counts of each state slightly deviates from frame to frame, which also varies the final calculated count rates. This will be analyzed more in detail in the next section.

We analyze the mean count value λ more in detail by applying different switching frequencies to the current controller. Using the matrix scheme Eq. 5.8 the detector count vector $c_{t_{clk},X}$ can be calculated. As depicted in Fig. 5.21, detector 3 has for all measurement time durations the highest counts and detector 4 the lowest. For all channels, the linear dependency of the counts on the measurement time t_{clk}, which was already demonstrated with one single channel without multiplexing approach (see Fig. 5.16), can also be seen in the CDMA case. The slight deviations from the linear fit result from the variations in the optical flux of the laser diode.

The strength and composition of the count spread in CDMA measurements needs further analysis.

The aim to get spread-less results of the CDMA coded data leads to the following conditions:

1. The sequences of bias switching follow the Walsh functions correctly and with high precision;

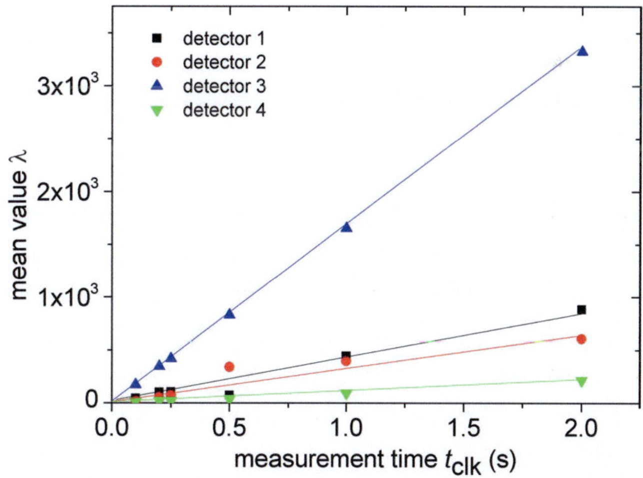

Figure 5.21: Linear dependence of the extracted counts of a CDMA measurement on the switching time t_{clk}. The absolute value is proportional to the momentary photon flux on each detector element. [HWE$^+$12]

2. The values of the measured count rates of all detectors are constant during T_{frame} because the vector c in Eq. 5.7 is assumed to be time invariant during one Walsh cycle;

3. The standard deviation of $c_{t_{clk}}$ is zero;

Condition 1 is fulfilled by switching the bias currents with the correct coding sequence. The precision, of course, is given by the electronics and the quality of the trigger signal. Condition 2 cannot be fulfilled, since in an imaging application, the incident photon flux is expected to have a certain dynamic. Each imaging frame typically depicts the progress in time and delivers a change in the mean count rate. That means, the CDMA could average photon fluctuations that appear during one frame time T_{frame}, but is not suitable to compensate a monotone drift of the photon flux during T_{frame}.

Even if the variation of the photon flux is almost constant during T_{frame}, each count measurement has a certain dispersion. This deviation contradicts to condition 3 and will downgrade the orthogonality. One can summarize that a CDMA measurement is not possible without spread in the extracted data. For operation it is important, how the spread could be reduced.

To verify the correctness of the counts of the CDMA measurement, three experimental situations are analyzed, exemplary. For each experiment a different photon rate is applied to the chip, leading to a variation in the counts of each pixel and a variation in the overall number of counts registered by the RSFQ electronics. In Fig. 5.22, the color of the columns represents one of the four detectors. The solid columns depict the mean value λ of the extracted counts during t_{clk} based on the statistical information of several tens of frame cycles. The striped columns represent the standard deviation σ from mean value λ.

The used measurement setup cannot be calibrated to get the real photon number passing each detector element to calculate the detection efficiency. So, a single detector sequence is sequentially measured for all four detectors as a reference for the CDMA measurement to verify the mean number of counts. This measurement is equal to a single-channel experiment. The deviation from this reference count number is marked as error bar on each λ column. The error bars shows that the measured mean values λ of the CDMA readout are, as expected from the theoretical predictions, very close to a time-gated single-channel measurement. The size of the error bars let conclude that the deviation from a single-channel measurement is marginal and not systematically shifted. It is mainly produced by the already mentioned slight variation in the photon flux during the measurement and the influences of the photon distribution.

Typical CDMA multiplexers use the typical enlargement of measurement time to increase the signal-to-noise-ratio (SNR) leading to a reduction of the noise floor on the output signal. However, the inaccuracy in the count measurements is not an uncorrelated white noise.

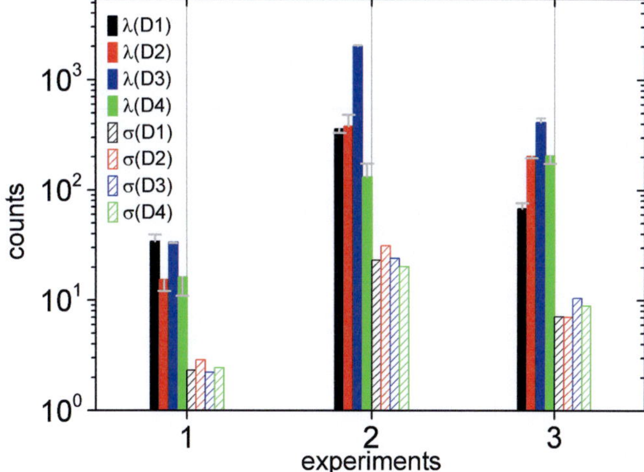

Figure 5.22: Mean value λ and standard deviation σ of orthogonal sequencing measurements for different photon flux conditions. The error bars mark the deviation from a single pixel measurement. [HWE$^+$12]

$$\sigma_{tot} = \sqrt{\sigma_1^2 + \sigma_2^2 + \sigma_3^2 + \cdots} \tag{5.10}$$

The standard deviations σ_X in Eq. 5.10, which correspond to the standard deviations of each detector element based on the active time $n \cdot t_{clk}$, are caused by the detector response itself and therefore correlated to the count values. Moreover, the summation of σ_X scales with the number of detectors (see also Fig. 5.19 and Eq. 5.14).

Figure 5.22 demonstrates, as well, how this influences the accuracy. For all three experiments the standard deviation of the mean counts is calculated. We can see for different illumination conditions that the standard deviations for all detector elements are on the same level. One can conclude from this measurement, that a CDMA system produces an accuracy level, which leads to similar standard deviations σ for all detector elements independent of the individual current count rate.

This has a logical reason because the counts of each detector element is always somewhere inserted in the extraction calculations of Eq. 5.8 of all the other pixels with either positive or negative sign, which leads to a geometric summation of the uncorrelated standard deviations of all detector elements.

Again, a complete pixel matrix is analyzed. The data bases on the counts vs. measurement times t_{clk} measurement from Fig. 5.21. In this case, the mean count rate λ and the standard deviation of the mean count rate σ is calculated for each measurement time t_{clk}(see Fig. 5.23).

The mean count rates are obviously almost identical for all measurement times t_{clk}. A slight spread in the mean value can be seen, which results from the already mentioned flux

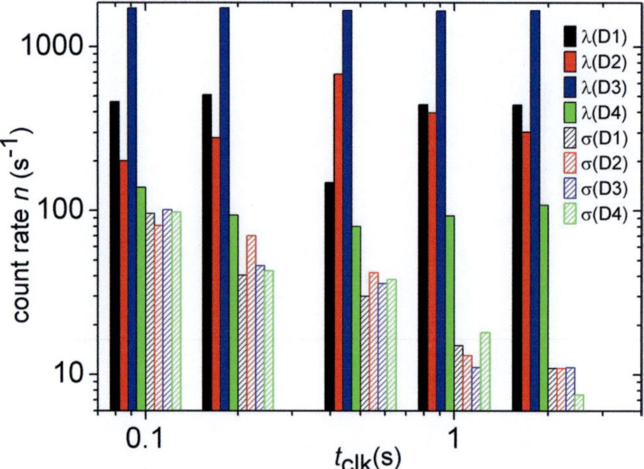

Figure 5.23: Decoded count rates for different durations of the clock time t_{clk} during a CDMA frame. The filled columns are the mean values of counts λ of the four pixels. The dashed columns are the dedicated standard deviations σ. [HWE$^+$12]

variation during the measurement (especially detector 1 and 2 at $t_{clk} = 0.5$s). Using longer measurement times t_{clk} the standard deviation decreases, which fits to the expectations of the Poisson distribution and the analysis of the single-pixel measurements of Fig.5.16. One can again recognize that as in the experiment before the standard deviations of all pixels are almost identical at a certain level σ_{tot} for each time t_{clk}.

One can conclude that a CDMA measurement of the count rate leads to a equalization of σ for all pixels with the final standard deviation σ_{tot}. Moreover, a longer measurement time or a higher detection efficiency, respectively, would improve this common accuracy level σ_{tot}.

5.3.7 Comparison between CDMA and TDMA

The analysis of the count accuracy of a CDMA measurement should be compared in relation to a time-division multiplexing access (TDMA) measurement to evaluate the advantages and disadvantages of both methods. A TDMA measurement means the sequential measurement of $2n$ detector pixels each with the measurement time $T_{frame}/2n$

The "on"-time of each pixel is in the CDMA case $T_{frame}/2$ and is independent of the number of pixels. Compared to a TDMA measurement, where the measurement time is inversely related to the number of pixels, this is an enlargement of the "on"-time of all detector elements and would increase the accuracy according to the results in section 5.3.4. However, merging several detector signals to one output leads to a cross talk (see Fig. 5.19), which leads to a degradation of the orthogonality.

Engert et al. demonstrated that the resolution limit of the CDMA can be traced back to a binomial distribution [EWH$^+$13]: In a simplified consideration, all pixels can be arranged in two types. Each pixel is either a strong counting pixel with the mean count rate μ_s or a weak counting pixel with the counting rate μ_w. The photon count rates are assumed to follow a Gaussian distribution $N(\mu_i, \sigma_2)$ (see Fig. 5.17). The mean value μ of the Gaussian distribution is identical to the mean value λ of the Poisson distribution in our experiments. A weak counting pixel μ_w can never be resolved by a CDMA approach with a lower variance σ_w^2 than with a TDMA approach, since the variance of a TDMA measured weak counting pixel is always $\sigma^2 = n_p \cdot \sigma_w^2 < n \cdot \sigma_w^2 + X$. X is the additional variance part, if one of the n_p pixels is a strong counting pixel.

A different situation is given for the strong counting pixel:

$$n_p \cdot \sigma_s^2 > 2(n-m)\sigma_w^2 + 2m\sigma_s^2. \tag{5.11}$$

n_p is the number of pixels, m_p is the number of strong counting pixels and $(n_p - m_p)$ is the number of weak counting pixels. The equation claims that the resolution of the common variance of the pixels resolved by a CDMA approach has to be smaller than the variance of a strong counting pixels in a TDMA approach with the dedicated time reduction factor $1/n$. This can be formed to:

$$\overline{\mu} < \frac{n_p - 2m_p}{2n_p - m_p} \tag{5.12}$$

with $\overline{\mu} = \frac{\sigma_w^2}{\sigma_s^2} = \frac{\mu_w}{\mu_s}$.

It can be concluded from Ineq. 5.12 that the strong count rates must be at least twice as big as the weak count rates, which means $\overline{\mu} < 1/2$. Moreover, if one solves Ineq. 5.12 for m_p, the number of strong counting pixels, one gets:

$$m_p < \frac{n_p - 2n_p\overline{\mu}}{2 - 2\overline{\mu}} \tag{5.13}$$

From Eq. 5.13 it can also be concluded that no more than half of the detectors may be strong pixels to deliver better results than a TDMA approach.

For the decision if CDMA is more suitable than TDMA, it is important to know, what the focus of the application is. If the number of counts of a strong counting pixel should be resolved with better resolution, the CDMA method is beneficial. If the weak count pixel should be resolved with higher accuracy than the mean deviation, the TDMA method has to be preferred. So, it is a question of the resolution of the imaging contrast, which method is more suitable.

The common standard deviation level σ_{tot} is a limitation for the absolute resolvability of weak count rates. It restricts the dynamic range of possible count numbers because detector elements with count numbers lower than the common standard deviation level can not be extracted correctly any more for a fixed time t_{clk}, as described in the following condition:

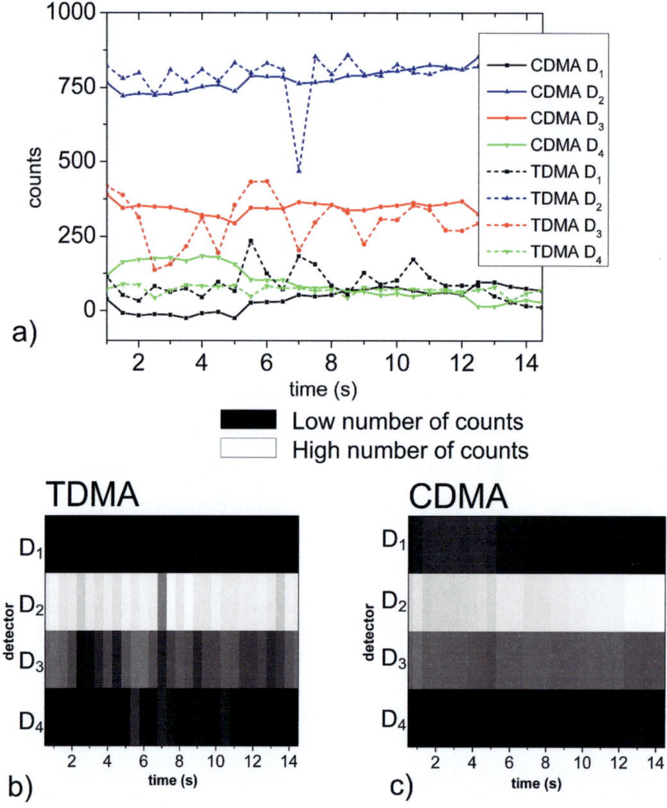

Figure 5.24: a) CDMA and TDMA counts over time directly compared. b) Grey-scale imaging of the TDMA measurement versus time. c) Grey-scale imaging of the CDMA measurement versus time.

$$\mu_X > \sqrt{\sigma_{tot}^2} \longrightarrow DE_X \cdot n_{ph} \cdot nt_{clk} > \sqrt{\sigma_{D1}^2 + \sigma_{D2}^2 + \cdots} \qquad (5.14)$$

We prove this with a combined measurement of a CDMA and a TDMA approach. Both types of bias switching sequences are directly applied after each other to the setup without dead time in between. The frame time T_{frame} is selected to $250ms$ and 30 frames are measured. The data of all four pixels is extracted and plotted in Fig. 5.24.

In Fig. 5.24 a) the counts per frame over time t are plotted. The solid lines represent the CDMA data and the dashed lines the TDMA data. It can be recognized that for the pixels with higher counts (D_2, D_3) the CDMA provides better accuracy than the TDMA. For the pixels with weak counts (D_1, D_4) the spread over time seems to be similar in this graph, however, one can see that the course of the CDMA is shifted compared to the course of the TDMA and the CDMA data reaches partly even negative values. This general shift defines a deviation, as well, and fits to the mathematical prediction and the negative count values indicate that the count rate is close to the absolute resolvability of the current conditions. In Fig. 5.24 b) and c), the data of a) is evaluated in a "grey-scale" picture. "White" corresponds to a strong count number and "black" to a weak count number. One can see for all pixels that the CDMA delivers a smother course of the counts per frame over time. The CDMA approach is more suitable to average these fluctuations. Although the weak count pixels have a deviation from the expected mean value, the deviation of the weak count pixel and the negative count numbers do not really affect the optical interpretation. Nevertheless, the results of the TDMA are more exact for weak counts.

5.3.8 Summary and outlook

Imaging with a very high sensitivity requires SNSPDs and a suitable multiplexing concepts. In this section, it is demonstrated, how a signal conversion between an SNSPD pulse and a RSFQ circuit can be realized. In a first experiment the interface between the SNSPD and the RSFQ electronics is demonstrated with regard to a loss-less signal conversion. In a further experiment it is demonstrated with a 4-pixel array, how a multiplexing concept can be applied. The concept uses a combination of a superconducting RSFQ-based merger electronics and a multiplexing scheme derived from communication technique to enable local resolution of photon rates. It is compared and analyzed in which situation time-division multiple access (TDMA) and code-division multiple access (CDMA) provide a better resolution of photon rates by decoding a complete imaging frame. It could be shown, that a CDMA readout leads to better accuracy of the measured rate for the pixels with strong count rate. The pixels with weak count rates have a lower accuracy with CDMA in the evaluated count rate than with TDMA. However, this larger deviation of counts for the weak counting pixels can often be neglected in an optical interpretation of the final imaging picture.

The main challenge of the multiplexing approach is that the increase of the pixel number always adulterate the accuracy of a photon-rate measurement, especially if the time span of one imaging frame is fixed. In communication theory, for CDMA, methods like successive interference cancellation [128] or whitening concepts [129] are provided. Especially whitening procedures, which try to re-orthogonalize the measured data are interesting to improve the accuracy of the photon rate measurement in a post-process although keeping the measurement time constant.
Alternatively, one can think in connection with the signal conversion concept presented in [OHT$^+$13] of a multiplexing based on an individually coded stimulus of the input comparator, which can be extracted by later correlation arithmetic.

5.4 Real-time multipixel readout of SNSPDs with time-tagged detector arrays

In the last subchapter a concept for real-time framing of an SNSPD multipixel matrix is discussed. The core task of this concept was the evaluation of the average count rate in a certain time frame. The evaluation of the average count rate is already a statistical information based on many measured photon events. An evaluation of the information given by a single photon is not possible with this concept.

In quantum cryptography, quantum optics and spectroscopy the single-photon event needs to be measured in real-time and directly evaluated. Moreover, a multipixel system should be continuously observed to prevent a registration error of count events. Although generally applications gain from a higher pixel number, it is not compulsorily required to reach large numbers of pixels for the types of applications described above, e.g. in the field of spectroscopy, the materials are often analyzed by a line of detectors in connection with line scanning mechanics [73]. Another application could be a row of detectors in combination with a spectral grating, which offers the possibility to resolve the spectral energy and photon number by lateral analysis.

For all these applications, a line of detectors is required, whose overall number is in the range of several tens of pixels. Therefore, the challenge is to develop a multiplexing concept, which requires only low system complexity to readout a moderate number of detectors. The following work is in part published in [HAI$^+$13].

5.4.1 Time-tagged multiplexing

For real-time single-photon multipixel readout, a concept is introduced, which bases on a time-tagged multiplexing (TTM) of an SNSPD array. This concept allows to assign the count event to the correspondent pixel in real-time.

The basic idea is schematically demonstrated in Fig. 5.25. Several SNSPDs are designed on one chip. The detector elements are connected by superconducting delay lines, which leads to a sequenced order of all detector elements. The detector chain is connected on one side to the readout line, where it is also connected to a single bias line, and is grounded on the opposite side. The delay lines between the detector elements add an individual delay time to the detector response pulse which gives each response pulse a specific arrival time relative to the timing of the photon event. This concept enables a low cryogenic system complexity and a single bias line to supply all detector elements because of the sequenced order of the detectors. Moreover, only one readout line is required.

The principle of operation is derived from the basic detection mechanism model. Typically, a detector is continuously biased with a bias current I_B close to I_C. Consequently, I_B

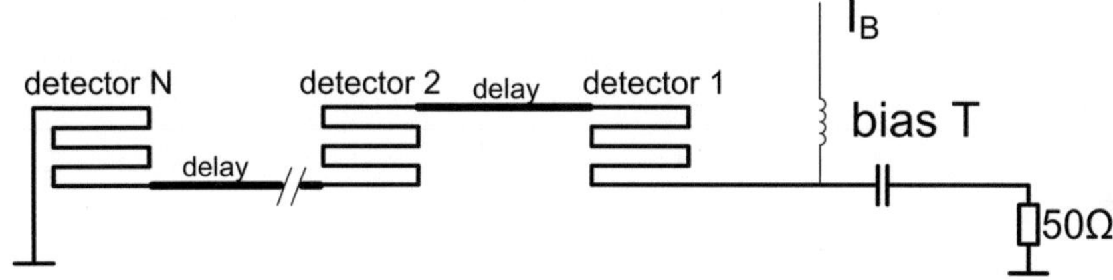

Figure 5.25: Concept of time-tagging of several SNSPD detector elements by placing a delay line in between. The scheme requires only one bias and one readout line. [HAI$^+$13]

simultaneously flows through all serially connected detector elements and the corresponding interjacent delay lines. Therefore, all detectors are activated at the same time. After an absorbed photon generated a hot spot in one of the nanowire detectors, a normal conducting belt arises across a small part of the meander line. The advantage of the detection principle is that the readout line is indifferent to the position in the detector path where the normalconducting belt occurs. All pulse events, which are generated by the detector elements, arrive at the readout electronics, however, individually delayed. The single condition is that the system works in the single-photon regime because, after a photon event, the bias current collapses along the complete detector chain, deactivating all other pixels until the pulse response relaxes and I_B returns back in the detector path. Consequently, this method does not have photon number resolving capability.

To measure the arrival time of a pulse response a defined reference point of the photon event is required. A simple way to define this is to use a pulsed-laser excitation. Most of the experiments in spectroscopy (e.g. those, which excite biological or chemical samples in form of scattered light or fluorescence) or quantum optics are already excited with a pulsed laser system. The used laser pulse width is, depending on the application, in the range of pico- or femtoseconds, with repetition rates up to 100 MHz [TGS$^+$12]. To evaluate the time signature of each detector element, the time-correlated single-photon counting (TCSPC) method [73] is used (see section 3.7). The TCSPC measures the time difference between the single-photon detector response and the reference pulse from the pulsed-laser source. One has to note that the time-tagged concept only works for applications, where the temporal length of the radiation effect under investigation is shorter than one delay of the detector chain. Moreover, the TCSPC technique can only be applied, if the probability for the occurrence of a count during one excitation pulse is $\ll 1$. Therefore, it has to be guaranteed that the occurrence for a multiphoton event is negligible.

5.4.2 Definition of delay time

The minimum acceptable duration of the delay is given by the transit time spread (TTS, also called detector jitter t_j) of a single SNSPD. If the TTS of the detector is broader than the delay time of one delay element, the time resolvability of the TTM is destroyed. A suitable value is a delay time with at least twice the TTS of the single-detector answer.

The delay time τ_d per line length is defined by the group velocity v_{ph} of the detector pulse. In superconductors, the delay time can be calculated based on the overall inductance $L_{delay} = L'_{delay} \cdot l = L_{geo} + L_{kin}$ to:

$$\tau_d = \frac{l}{v_{ph}} = l \cdot \left(\frac{1}{\sqrt{L'_{delay} \cdot C'}} \right)^{-1} = l \cdot \sqrt{L'_{delay} \cdot C'} \tag{5.15}$$

with l the delay line length and L'_{delay} and C' the inductance and capacitance per unit length. In thin and narrow superconductors, the kinetic inductance L_{kin} has a dominating influence on the overall inductance L_{delay} [130], [20], [51]. In the normal conducting case the propagation delay time ($L_{delay} = L_{geo}$) is given by:

$$\tau_{d,n} = l \cdot v_{ph,\,n}^{-1} = l \cdot \left(\frac{c_0}{\sqrt{L' \cdot C'}} \right)^{-1} = l \cdot \left(\frac{c_0}{\sqrt{\mu_r \cdot \varepsilon_{r,eff}}} \right)^{-1} \tag{5.16}$$

The delay time $\tau_{d,\,n}$ is only dependent on the length l of the delay line and the effective permittivity $\varepsilon_{r,eff}$ (for the used detector environment, one can assume $\mu_r = 1$). In the superconducting case, where the magnetic field can penetrate into the superconductor on the length λ_L, the propagation velocity $v_{ph,\,s}$ slows down and the delay time $\tau_{d,\,s}$ increases. Especially, in very thin films, where the film thickness $d \ll \lambda_L$, this might lead to a noticeable increase of the propagation time [131]:

$$v_{ph,s} = \frac{v_{ph,n}}{[1 + \lambda_{L,2}/h + (\lambda_{L,1}/h)\coth(d/\lambda_{L,1})]^{\frac{1}{2}}} \tag{5.17}$$

h is the substrate thickness and d the superconducting film thickness. The superconducting velocity $v_{ph,\,s}$ is decreased compared to the normal conducting $v_{ph,n}$ due to the denominator in Eq. 5.17. The London penetration depth λ_L is split into two parts $\lambda_{L,1}$ and $\lambda_{L,2}$. $\lambda_{L,1}$ is the value of the superconducting transmission line and $\lambda_{L,2}$ the value of the material of the GND plane, if also superconducting. Since the plate on the bottom of the detector is from normal conducting brass/copper, respectively silver paint, the summand with $\lambda_{L,2}$ can be set zero.

5.4.3 Design of a multipixel detector

All SNSPDs are made from a 5 nm thick NbN film. The NbN films are completely patterned by electron-beam lithography and ion milling as described in section 4.1.2 and 4.1.2. For analysis several chains of detectors are fabricated:

- a 2 detector chip with different delay lengths,
- a 4 detector chip

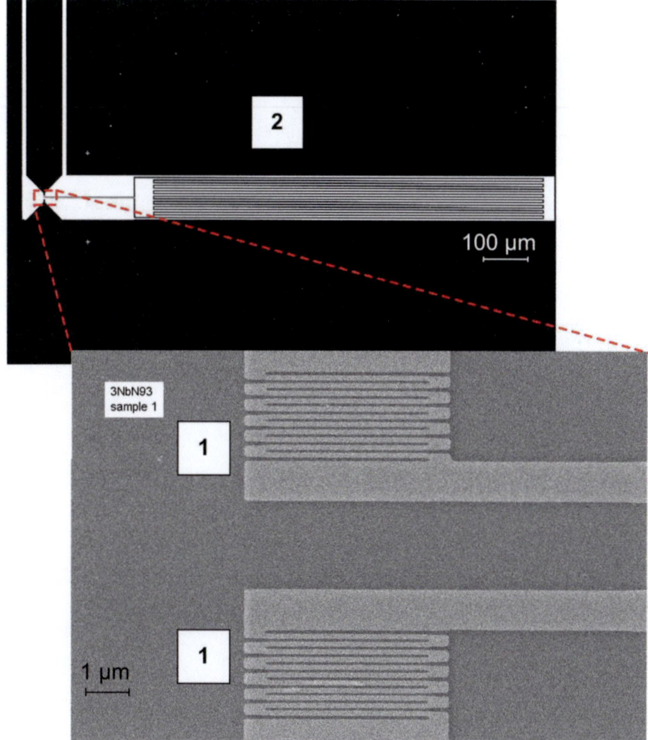

Figure 5.26: (1) Two detector pixels. (2) Connection by a delay line embedded in a coplanar read out. The distance between the pixels is 4 µm. [HAI$^+$13]

Two-pixel detector

For the first proof of principle, two detector elements (size: 2x4 µm, linewidth $w_{SNSPD} = 100$ nm) are designed and fabricated in two versions, each with different delay length in between. The lateral distance between the two SNSPDs elements is 4 µm. In Fig. 5.26, the elements (2) are coupled on one side by a delay line made from the same NbN film as the detector. Outside of the magnified view (Fig. 5.26 (2)), the chain of detector elements is connected to GND and the coplanar waveguide.

The width of the delay line is selected to be $w_{delay} = 1$ µm, as a trade-off between a low inductance, maintaining the superconducting requirements and a reduction of the required space on the chip. Additionally, the delay lines are meandered (see Fig. 5.26 (2)). The critical current $I_{C,\,delay}$ of the delay line is roughly 10 times higher than the critical current $I_{C,\,det}$ of the detector line ($w_{det} \approx 100$ nm) to ensure that the delay line does not detect photons itself but solely acts as a delay for the traveling detector pulses.

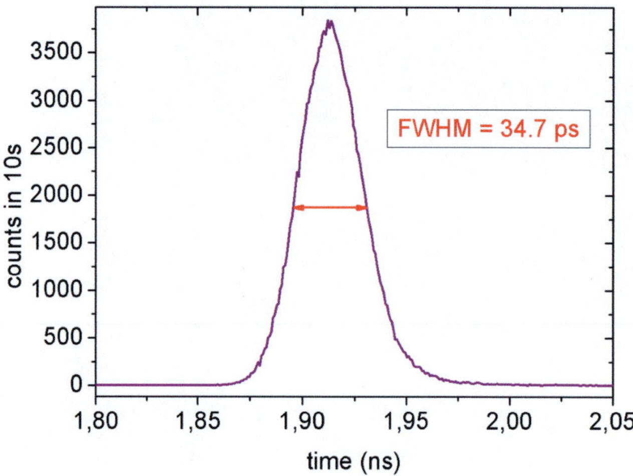

Figure 5.27: Typical jitter of an SNSPD.

Fig. 5.27 shows a typical TTS measurement of a single-pixel detector with the identical film thickness and meander line width as used in the 2-pixel experiment, which reaches values in the range of 35 ps. Details of this experiment are published in [TGS$^+$12]. An acceptable delay for the time-tagged measurement is in the range of > 70 ps (twice the typical TTS).

With common numbers for NbN ($\lambda_L = 400$ nm) [47] and typical geometric numbers of the detectors (substrate thickness $s = 300$ μm, film thickness $d = 5$ nm), one can estimate based on Eq. 5.17 an enlarged propagation time in the superconductor of around $\tau_{d,s} = 2.5 \cdot \tau_{d,n}$. However, the substrate thickness is larger by a factor of $60,000$ than the film thickness and by a factor 300 larger than the line width. Therefore, one can assume that probably a big part of the microwave signal is located as electromagnetic field in the dielectric medium, traveling with the dielectric propagation delay time τ_n defined by Eq. 5.16.

The permeability μ_r is supposed to be 1. The permittivity $\varepsilon_{r,eff}$ can be approximated by $\varepsilon_{r,eff} = (1 + \varepsilon_{r,sapphire})/2$, which takes into account the geometric layout. This can be considered as a kind of grounded coplanar waveguide because the delay line is completely surrounded by a superconducting ground plane and has a metallic copper layer on the back side of the substrate. In consideration of the minimum required delay time estimated above, the delay line lengths are selected to be $l_{d1} = 11$ mm ("short delay detector" - SDD) and $l_{d2} = 20$ mm ("long delay detector" - LDD). The dedicated normal conducting delay times of these two delay lines are calculated to $\tau_{d1} = 86$ ps and $\tau_{d2} = 155$ ps by Eq. 5.16.

For both delay lengths, a series of samples with identical design parameters is fabricated. For all samples the critical current I_C, and the critical temperature T_C are measured. For

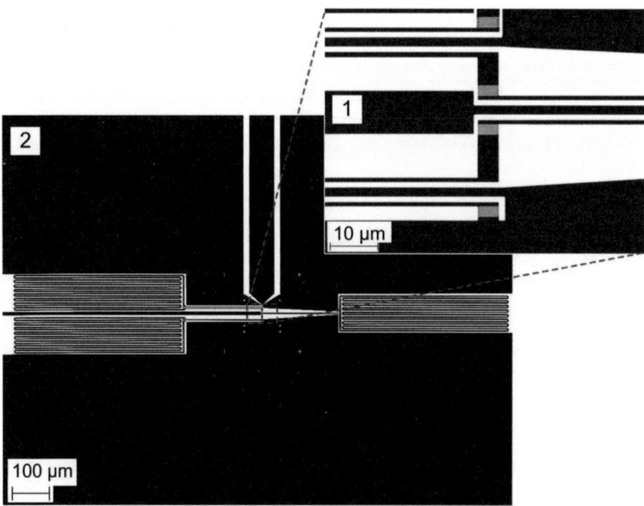

Figure 5.28: (1) Magnified chain of four detector elements. The lines at the meanders lead to the delay line structure. (2) Coplanar readout connected to the 4-pixel chain with three meandered delay lines embedded.

further detailed optical characterization, the two devices (one SDD and one LDD) with the highest values in I_C (≈ 40 µA) and T_C (≈ 12 K) have been selected.

Four-pixel detector

The fabrication process of the four-pixel design is identical to the two-pixel design. The four pixels are separated by three delay lines in between, which are put on both sides of the readout axis, see Fig. 5.28. The delay line has a width of 2 µm and a length of $l_d = 7.3$ mm, which leads to $\tau_d = 57$ ps. The GND layer is optimized, by guiding a GND line along the signal lines to improve the coplanar behavior. The lateral distance between the pixels varies between 4 and 12 µm due to restrictions of the delay leads.

5.4.4 Time-tagged measurement setup

For the measurements, the dip-stick cryostat from chapter 3 is used. The cryostat is cooled down to 4.2 K by liquid helium. The input radiation stems from the 1550 nm fiber-based fs-laser with 100 MHz repetition rate. The detector chain is mounted on the copper block and biased by the cryogenic biasT. The output signal is subsequently amplified with 53 dB gain by three room-temperature amplifiers with an effective bandwidth of 1.7 GHz. Directly at the cryostat output, there is an additional 3 dB attenuator for better rf matching.

The TCSPC measurements are done with the TCSPC electronics from Becker & Hickl (Simple-Tau-130-EM-DX). The SNSPD pulse amplitudes are adjusted by applying a 6 dB attenuation to adapt them to the maximum input range of the TCSPC electronics. The time reference for the TCSPC measurement is the internal reference output of the fs-laser. The pulse shape measurements are done with a 32 GHz oscilloscope.

5.4.5 "Proof of principle" of time-tagged multiplexing with two-pixel SNSPD

An important factor of such multipixel system is the detection efficiency of the detector pixels. It is not possible to measure the detection efficiency for each detector element separately because the detector elements are positioned too close to each other on the chip for individual illumination and a separate biasing is not possible due to the concept of using one common bias line. A measurement of the common detection efficiency of both chips at 4.2 K delivered values of $DE = 2$ % in the optical range monotonically decreasing in the infrared range, which is comparable to a single-pixel detector [HRI$^+$10b] and a proof of a homogeneous DE on the chip. That means, the sensitivity of the detector meanders is not suppressed by the presence of the extended and sequenced detector path.

Since the single pixel in Fig. 3.20 generates one Gaussian peak in a TCSPC measurement, two delayed detector elements should deliver two Gaussian peaks with a temporal distance τ_d corresponding to the delay line length l_d. Figs. 5.29 and 5.30 show the results of the short-delay (SDD) and long-delay (LDD) detectors, both measured at bias current $I_B = 0.9 \cdot I_C$.
As expected, the measurements show two Gaussian peaks in the histogram with different distances in between. The t_j of the single pixels is broader than 35 ps, which is due to the non-optimized cabling lengths and the use of the multimode fiber.

Varying the photon flux on the detector elements (the beam spot of the multimode fiber has a mode profile, which can be used for varying the photon rate), the amplitudes of the Gaussian peaks vary. However, τ_d stays independent of the incident power level.

Conversely, it is found that τ_d is dependent on the discriminator level of the TCSPC electronics of the SNSPD input (the meaning of the discriminator level varies slightly from typical considerations because the TCSPC electronics uses a constant fraction discriminator (CFD) (see [78] and section 3.7)). Figure 5.31 shows the delay time τ_d of the detector chain for different discriminator levels of the TCSPC electronics. Obviously, a strong dependence on the discriminator level for the LDD chip can be observed. The SDD chip is independent with only small deviations which originates in the accuracy of the measurements. One reason might be that the pulse of the LDD sample response has stronger variations in the pulse amplitude than that of the SDD sample. The TCSPC electronics is able to compensate an amplitude jitter by employing the CFD at the inputs. Nevertheless, the parameters of the compensation are only valid for a narrow amplitude range, which might lead to spread in cases where the amplitudes behave quite different. The maximum delay time in Fig. 5.31 is measured at the point where the pulse has the steepest slope. The delay times τ_d (94 and

156 ps) fit well to the calculated delay times based on the geometric inductance. Consequently, it seems convincing that the influence of the kinetic inductance on the propagation velocity is negligible under these geometric conditions.

Using the time-tagged readout in an application, the real-time assigning of the corresponding pixel during the measurement can directly be done with the measured time-tag. A time-tag belongs either to the time interval of the right or the left Gaussian peak in Figs. 5.29 and 5.30, which decides about the local origin of the pulse event. The small overlap between the pixels, however, is an undefined zone. Therefore, a real-time assigning needs a quite good separation of the Gaussian peaks, which can be generally reached either by increasing the delay length or shrinking the TTS of each detector element. The discriminator level dependence of the delay time could also be an operation parameter to modify the delay time, but needs better understanding.

From the TCSPC measurement, estimation about the scalability of the pixel number can be done. Taking twice the t_j in Fig. 3.20 as minimum distance between two elements, which is $2 \cdot 35 \, \text{ps} = 70 \, \text{ps}$, and using a 100 MHz pulsed laser system, then up to $(1/100 \, \text{MHz})/70 \, \text{ps} = 142$ detector elements could be integrated in one array. A lower repetition frequency would increase the number of slots for detector elements.

5.4.6 Analysis of the response pulse of a two-pixel SNSPD

The understanding of the dependence of the delay time τ_d on the discriminator level requires a deeper analysis of the pulse amplitude.

The detector pulses of the SDD chip are recorded by a 32-GHz oscilloscope. Fig. 5.32 a) shows a plot of several measured pulses and the dedicated pulses of the laser reference output. The oscilloscope is triggered on the SNSPD channel on the falling edge of the

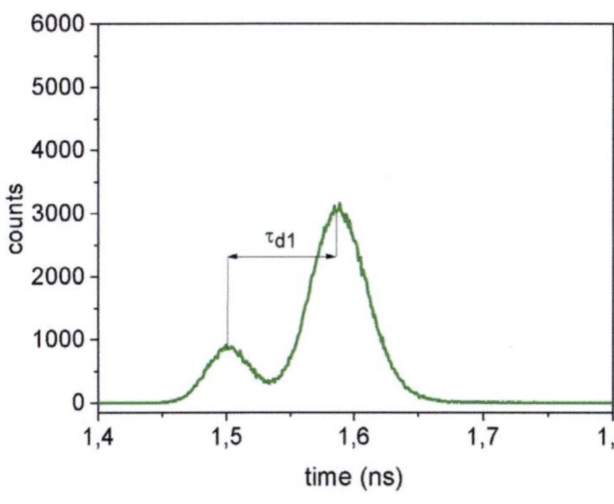

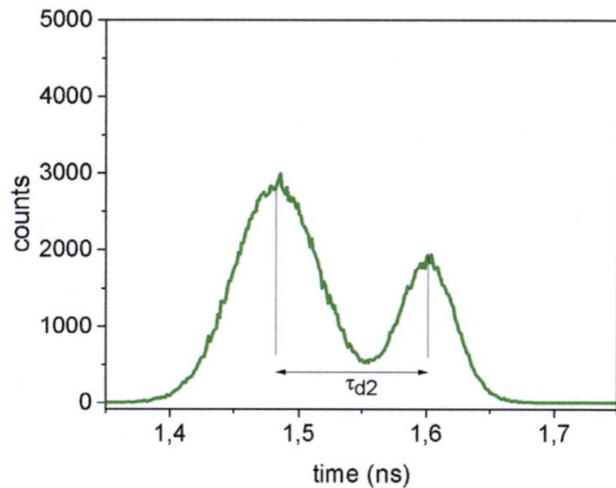

Figure 5.29: TTS measurement of the short delay detector (SDD) with $\tau_{d1} = 90$ ps. [HAI$^+$13]

Figure 5.30: TTS measurement of the long delay detector (LDD) with $\tau_{d2} = 119$ ps. [HAI$^+$13]

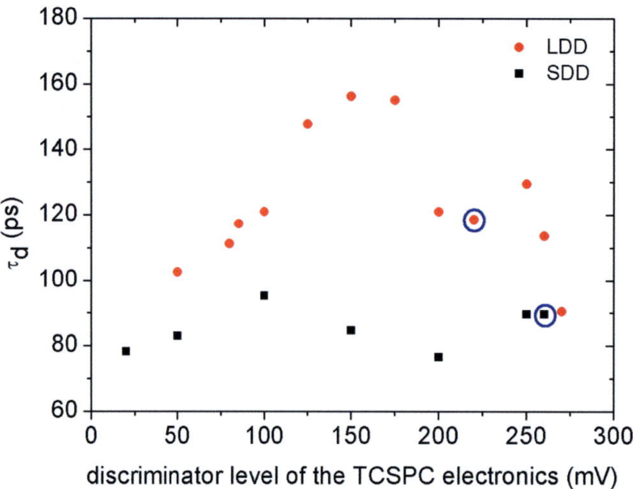

Figure 5.31: Dependence of the TTS on the discriminator level of the constant fraction discriminator input stage. The LDD detector shows a quite high dependence on the discriminator level and has a maximum. The SDD is almost constant vs. the discriminator level. The points marked with cycles belong to the measurement conditions of Fig. 5.29 and Fig. 5.30. [HAI$^+$13]

inverted SNSPD pulse. Therefore, the SNSPD pulses lie on each other and the time delay between the arriving SNSPD pulses is shifted to the reference pulses of the laser source. The laser reference pulses in Fig. 5.32 a) definitely show an accumulation at two defined times with a delay in between. The SNSPD pulses show two different slopes. One type looks like a standard SNSPD pulse and the other has a step in the front edge. Comparing each recorded pulse directly with the dedicated point of time of the reference pulse, one can determine that the standard pulse always belongs to the pixel which is located in the chain closer to the GND plane, and the pulse with the step belongs to the pixel which is closer to the readout line. Extracting the time delay between the reference pulses one gets for the SDD chip a delay $\tau_d = 164$ ps, which is longer than in the TCSPC measurements, but still smaller than predicted by the superconducting delay $\tau_{d,s}$ for the SDD chip. A measurement of several multipixel chips accords with this result.

The step in the pulse edge can be explained by considering the detector chain as an rf circuit. In the moment of the occurrence of the normalconducting belt, the current I_B is reduced in the detector element with a certain velocity of change and a pulse U_{det} is created by the "Induction law" $U_{det} = -L \cdot \frac{dI}{dt}$. This pulse U_{det} is positive due to the negative gradient in $\frac{dI}{dt}$. However, looking from the GND side into the chain, one can define I_B with a negative sign, as well and can expect that a pulse U_{-det} is created with negative amplitude traveling from the normalconducting belt along the delay line to the second pixel and the GND short. At the position of the second pixel the negative pulse U_{-det} is reflected on the common impedance $Z_L = 0\Omega + jL_{detector}$, which leads to a reflection factor $r \approx -1$ plus phase shift.

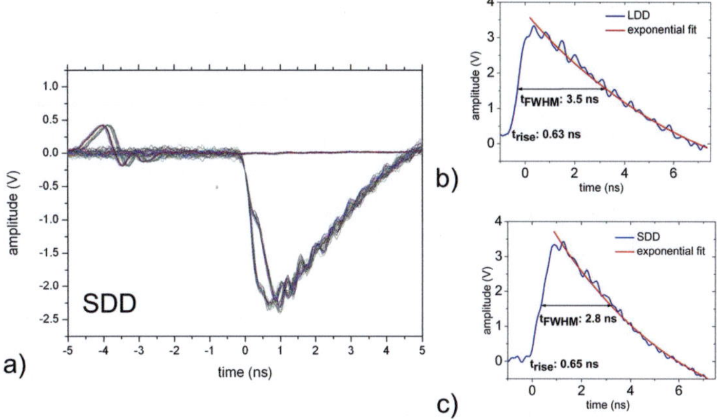

Figure 5.32: Oscilloscope measurement of a 2-pixel time-tagged array. a) Comparison of the two amplitude shapes of the two detector elements from one chip (SDD). The inversion of the pulse shapes results from the amplifiers. b) and c) Selected inverted response pulse shape of a LDD and a SDD chip for length analysis. The amplitude is inverted for better clarity. The exponential fits deliver a relaxation constant $\tau = 7.2$ ns (LDD) and $\tau = 5.5$ ns (SDD).

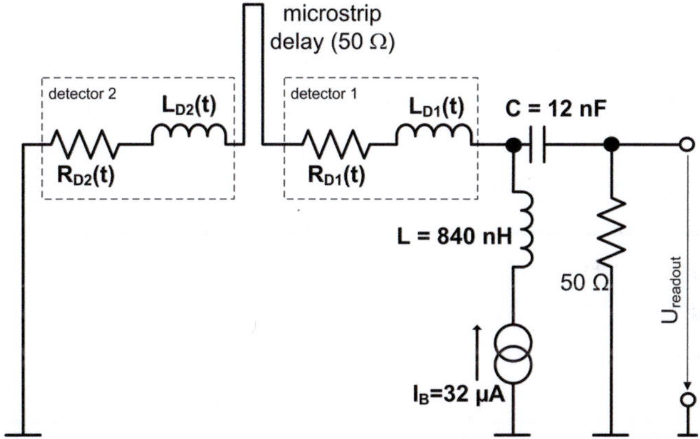

Figure 5.33: Model of the delay chain. Two equivalent circuit diagrams of the detector elements are in series with a delay line in between. The biasT components as used in the experiments are part of the simulation, as well.

The reflected positive pulse U_{r-det} travels back along the delay line and superposes after $t = 2 \cdot \tau_d$ the original pulse $U_{ges} = U_{det}(t) + U_{r-det}(t)$. The existence of the second negative pulse is shown in section 5.4.8.

The dynamics of the response signal of the chip can be verified with an rf circuit simulation of the complete delay structure with Agilent Advanced Design System (ADS):

The dedicated circuit structure is shown in Fig. 5.33. $R_{Dx}(t)$ is modeled as an asymmetric triangle function with a front slope of 100 ps and a decay of 700 ps based on the maximum relaxation times calculated by Eq. 2.8 with $\tau_{ep} = 7$ps, $\frac{c_e}{c_p} = 6$ and $\tau_{es} = 115$ps. The maximum value $R_{max}(t)$ is taken as fitting value. A suitable value for $L_{Dx}(t)$ is more complicated to define. The calculation of L_{kin} according to Eq. 2.15 deliver with $\lambda_L \approx 450$ nm [47], [9] $l = 7.3$ mm, $w = 100$ nm and $d = 5$ nm a value of $L_{kin,0} = 20$ nH. With Eq. 2.16 one can estimate the current dependent increase to $L_{kin,I_B} = 1.09 \cdot L_{kin,0} \approx 22$nH. The frequency dependence of the conductivity increases L_{kin} [130], as well. However, this increase can not be predicted exactly due to a unknown surface resistance $R_s(f)$ of the NbN film and the broad frequency spectrum of the SNSPD pulse. L_{geo} is neglected. Consequently, a $L_{Dx} = 40$ nH is assumed as stationary value, which fits to the values in [95], as well. Since L_{kin} depends on I_B and I_B decreases during the pulse event, L_{kin,I_B} is modeled dynamically according to Eq. 2.16. To account for the pulse shaping of the amplifier, as well, a bandpass is added, which has the same frequency range as the used amplifiers in the experiment. The gain of the amplifier is not included, but the simulated pulses are inverted for better comparability. In Fig. 5.34, it can be recognized in the simulation results both, the step in the pulse shape and the oscillations in the pulse decay which well agrees with the measurements. However, the reflected pulse overlays the main pulse in the simulation later in time and the oscillations are stronger. An explanation can be given: Although for the superconducting case, we respected that $L_{kin,SNSPD}(t)$ reduces with reducing the bias current I_B during the pulse event by ≈ 9 % according to Eq. 2.16 one has to consider, as well, that the normalconducting belt of the meander has no kinetic inductance according to the two-fluid model. Therefore, $L_{kin,SNSPD}(t)$ shrinks further. Furthermore, it is not clear, if the detector element which just guides the pulse signal through becomes normalconducting, as well. Since the current is redistributed in the detector path during a pulse event, it is more likely that the signal guiding detector element stays superconducting, however, with reduced $L_{kin}(t)$, as well. It is essential that the voltage pulse $U_{det} \propto \frac{dI}{dt}$. The slope $\frac{dI}{dt}$ is defined by the time constant $\tau_{el} = \frac{L(t)}{R_{det}(t)}$. Here, $R_{det}(t)$ describes the time-dependent evolution of the normalconducting belt. It can be concluded that in a superconducting consideration τ_{el} changes during the pulse evolution and $\frac{dI}{dt}$ follows this change. Therefore, the two pulses have different slope steepness. Though varying the bandpass frequencies in the simulation has shown that the steepness of the front edges of the pulses is already in a regime where it is influenced by the amplifier bandwidth. Although this is not an unexpected behavior, one should be aware that this leads to an equalization of the front edges in the steeper regions of the front slopes, which changes the trigger point, again.

In the 2-pixel measurement the oscillations of the pulse shape are only slightly distinctive compared to the simulation but agree better in amplitude for higher number of pixels (see section 5.4.7). They originate in the fact that the delays behave with decreasing length more and more as an oscillator with a standing wave depending on the delay length l_d and the additional inductive elements in the delay chain. Changing the delay length in the simulation verifies this behavior. However, $L_{kin,\,delay}$ is not considered in the simulation. In the measurement $L_{kin,\,delay}$ would definitely influence the standing wave evolution due to the

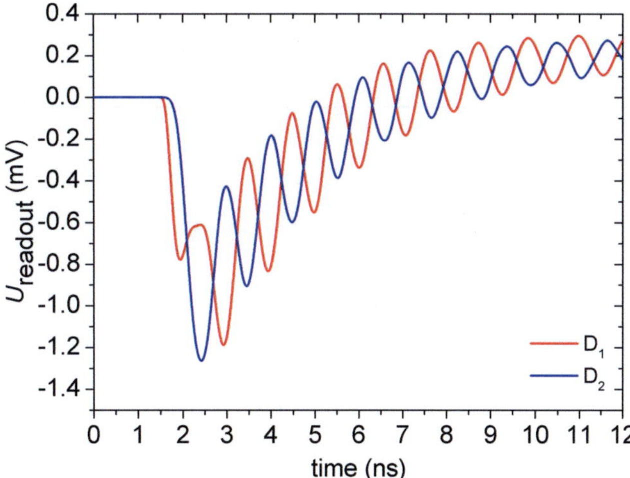

Figure 5.34: Simulated curves voltage versus time of the response pulses of a 2-pixel detector chain. The gain of amplifiers is not considered in the simulation. D_1 is the detector element directly connected to the readout line and D_2 is the detector element directly connected to the GND plane.

fact that this additional impedance is distributed along the delay line and not localized at the ends of the delays compared to the detector inductance, which probably attenuates the oscillations. It is also not clear in which way $R_{det}(t)$ attenuates the oscillation during its life-time. Moreover, a comparison has shown that the type of rf amplifier influences the strength of oscillations of the response pulses, as well, which can be seen in the measurements in section 5.4.8.

The different edge steepness of the two pulse types are the reason for the deviation in time between TCSPC measurement and oscilloscope measurement. Averaging several pulse shapes from each of the two pulses leads to the plot in Fig. 5.35. In the non-magnified view, the voltage curve between the start of the pulses and the point of time of the superposition with the reflected pulse – the step in the pulse edge – seem to be identical for both pulse types. Magnifying the pulses in this time span shows a different behavior. The pulse shape of the detector which is placed closer to the readout (detector 1) has a steeper increase of the amplitude at the beginning of the pulse evolution than the detector placed closer to the GND plane of the chip (detector 2). The inductance part of $\tau_{el,1}$, which defines the pulse evolution of detector 1 consists manly of the inductance $L_{kin,\ SNSPD,\ det1}$ in the considered time span. The inductance part of $\tau_{el,2}$ of detector 2 consists of $L_{kin,\ SNSPD,\ det2}$, and during the traveling to the readout line of L_{delay} and $L_{kin,\ SNSPD,\ det1}$, as well.

At the beginning of the pulse, $R_{det}(t)$ is still quite small. Consequently, $\tau_{el}(t)$ is in a time scale, in which the different inductances L of the both detector elements have a visible, respectively measurable influence on the amplitude curves. If $R_{det}(t)$ increases by several orders of magnitude with time, and L_{kin} reduces due to the reduced I_B, the different induc-

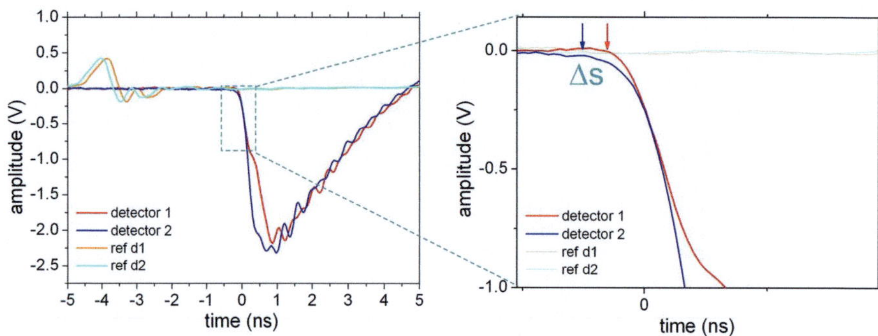

Figure 5.35: Averaged and magnified pulse shape of the two pixel pulses.

tances L are negligible in the dedicated resolvable time scale, which leads to similar voltage curves in the course of the pulse edge.

According to the magnified view in Fig. 5.35 the different time slopes produce a virtual shift ΔS of the actual moment of photon absorption. Since the oscilloscope needs a certain trigger point above the noise level of the measured channel, the oscilloscope always measures the common delay time $\tau_d + \Delta S$. The TCSPC electronics can use the CFD to reduce the influence of the different pulse shape evolutions, which leads to a measurement of only τ_d.

The electrical dynamics of the two pixels are defined by two different inductances L, which leads to two distinct amplitude curves after averaging the dedicated pulses, measured as virtual time shift. It can be supposed that in a single pixel meander different $\tau_{el}(t,L,R)$ should even exist in a smaller time scale, depending on the point of absorption on the meander and its dedicated inductance. This could give an explanation of the jitter t_j in an SNSPD. However, a proof would need further investigation.

The inductance of all components on the chip defines the full width of the SNSPD pulse, as well. Since the SNSPD response pulse width should be smaller than the period time of the pulsed-laser system, the pulse width is a further limit of the number of maximum detectors in one chain. In the insets of Fig. 5.32, the FWHM of the two pulses differ by ca. 0.7 ns in width. The lengths of the delays line differ by a factor of about two. Therefore, the SDD response ($t_{FWHM} = 2.8$ ns) can be split in two causes. 2 ns stem from the inductance of the detectors, which fits to the pulse width of a 5 nm thick single SNSPD [TGS$^+$12] whereas the additional 0.7 ns originate from the inductance of the delay line.

In the insets of Fig. 5.32 one can see that the relaxation times of the SNSPD pulses are more than ten times larger than the calculated delay times τ_d. Therefore, in contrary to the front edge of the pulses, the decay of the voltage curve can be considered as a lumped-element dynamics defining the time constant $\tau_{decay} = L_{total}/R_{total}$. Since $\tau_{decay} > \tau_\varepsilon$ from Eq. 2.8, $R_{det}(t)$ is already zero during the decay process. Therefore, one can set R_{total} to $R_0 = 50\ \Omega$, which is the remaining line impedance of the readout [20], and the total inductance $L_{total} =$

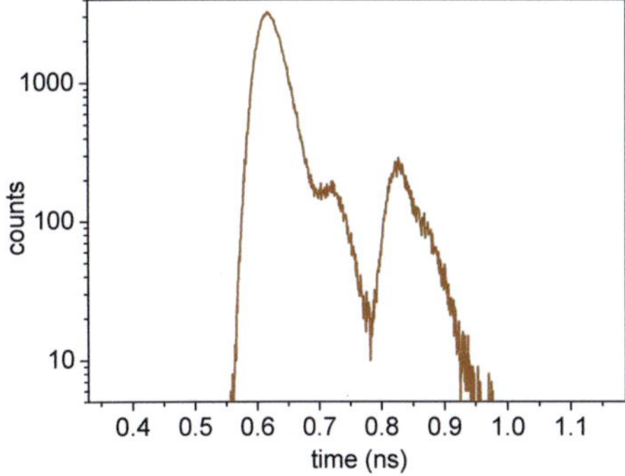

Figure 5.36: TTS measurement of the 4-pixel detector with TCSPC electronics

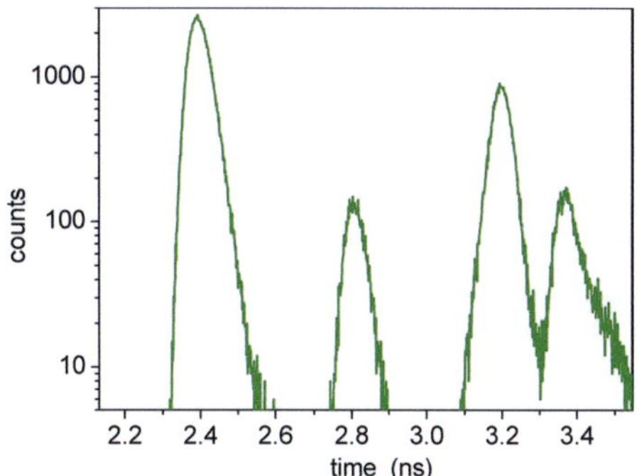

Figure 5.37: TTS measurement of the 4-pixel detector with TCSPC electronics and analog front end.

$L_{kin,\,SNSPD} + L_{delay}$. With the relaxation times $\tau_{decay,\,SDD}$ and $\tau_{decay,\,LDD}$, one can calculate L_{total} and finally, with the method from above, the common inductance of the two detector elements (each with $2x4\,\mu m^2$ in size) and the inductance of the delay of SDD: $L_{total,\,SDD} = 275$ nH, $L_{kin,\,SNSPD} = 190$ nH (for both detector elements together), $L_{delay,\,SDD} = 85$ nH. The value $L_{kin,\,SNSPD}$ of the detector elements is slightly larger than comparable values in literature [20], [27] and the estimated values based on film parameters, as described in the previous paragraphs. The enlargement of the extracted $L_{kin,\,SNSPD}$ is probably due to the additional pulse shaping by the limiting bandpass of the amplifier transfer function, which influences the fitting parameters.

Considering a scaling of the pixel number, the delay elements extend the pulse widths of the detector response, but not as much as by adding the further detector elements. Therefore, in future the focus should be on decreasing of L_{SNSPD}, which means mostly decreasing of $L_{kin,SNSPD}$. A slight improvement of several nH might be achieved by changing the superconducting film in direction of a lower λ_L or lower square resistance R_S. A reduction of L_{delay} can mostly be realized by enlarging the width and thickness of the delay line. Further improvement might be achieved, if the delay design is adjusted to use thicker and lower resistive films.

5.4.7 Time-tagged multiplexing with a four-pixel SNSPD.

Since the two-pixel time-tagging operates well, a chip with four pixels is taken to demonstrated the feasibility of this multiplexing concept for enlarged pixel numbers.

For the measurement, a multipixel chain is used with four detector elements, which requires three delay lines in between. The multipixel chip is measured with the TCSPC method (with two amplifiers) and the 32 GHz oscilloscope (with three amplifiers). The results are shown in Fig. 5.36 and Fig. 5.38.

The TCSPC histogram shows only three Gaussian distributed peaks, which are already partly overlayed. One can supposed that a fourth pixel is covered in the Gaussian curve of the third pixel, however, it can not be resolved anymore, as long as the count numbers, which define the height of the Gaussian functions strongly differ. The distance between the Gaussian peaks in the TCSPC measurement fits to the calculated delays of $\tau_d = 57$ ps. The dedicated oscilloscope plot in Fig. 5.38 shows both, the reference pulse from the pulsed laser source and the SNSPD response. The extracted delay times in the oscilloscope measurement are $\tau_d \approx 130$ ps. Each SNSPD pulse has a slightly different shape. One can recognize that the step in the front edge shifts along the edge depending on the pixel position and the strength of the oscillation is stronger with increasing total delay (detector 3 and detector 4). Although the trigger level of the oscilloscope measurement is set close to the GND level, the extracted delay time of the oscilloscope measurement is again broader than expected from calculations. An averaging of the pulses and magnified view of the pulse edges shows again the different slope evolutions of all detector elements, which leads to an additional virtual delay ΔS as in the two pixel case.

To improve the resolution of a measurement with TCSPC electronics, one can use this additional shift, to enable a full pixel resolution without longer delay lines. Therefore, in front of the TCSPC electronics the 10 GHz AFE module from section 3.7.2 is mounted, which allows a disabling of the corrections by the constant fraction discriminator, keeping the use of the TCSPC electronics. In Fig. 5.37, the new TCSPC measurement is shown. One can see four clear Gaussian pulses. The distance between the peaks is broader than given by the delay line, as expected, since the measurement is done at a discriminator level of the AFE, where the virtual shift ΔS already broadens the pulse delays. A variation of the discriminator level of the AFE has demonstrated this time broadening trend, as well.

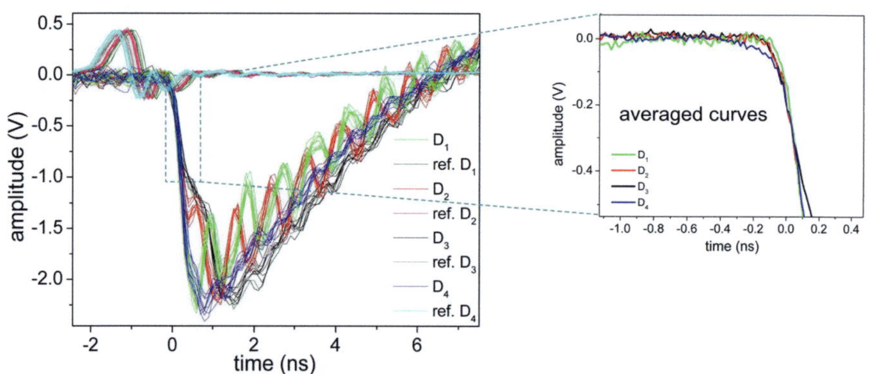

Figure 5.38: 4-pixel oscilloscope measurement. In the magnified view the slopes close to the GND level are shown.

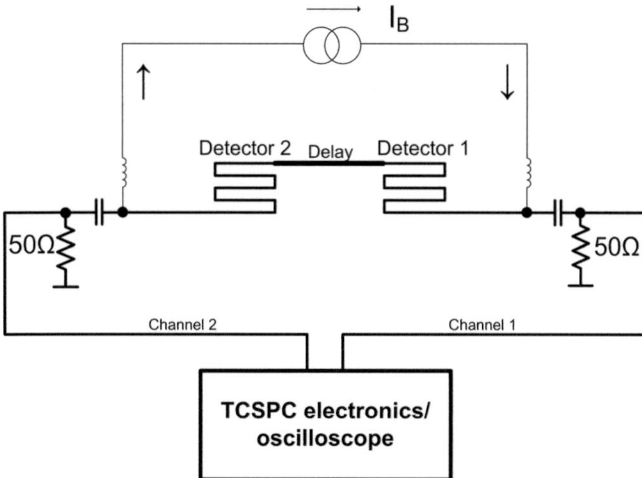

Figure 5.39: Readout scheme of a dual channel measurement.

An analysis of $L_{total} = L_{kin,\ SNSPD} + L_{delay}$ based on the pulse length for the 4-pixel chain is not possible because the pulse decay in the oscilloscope plot is mainly defined by the rf amplifiers and their lower bandpass cutoff frequency (the dedicated long negative tail is outside of the view of Fig. 5.38).

It can be concluded that the delay length can be kept shorter than required using the virtual delay ΔS, which leads to a decreasing L_{total}. Nevertheless $L_{kin,\ SNSPD}$ mainly defines L_{total}. Improvement can be achieved by increasing the cross section of the nanowire, which on the other side directly reduces the detection efficiency [HRI$^+$10b]. Better would be using a parallel stripe geometry ($L_{SNSPD,\ parallel} = L_{SNSPD}/n$ where n is the number of parallel stripes) [132].

5.4.8 Dual readout of time-tagged multiplexed SNSPDs for continuous wave radiation

Theoretically, one could assign each pulse response to the corresponding pixel during measurement without the need of a reference pulse by making a pattern analysis of the recorded pulse shape. However, this needs a pattern detection in connection with certain decoding time in a signal processing environment. A faster assigning was realized by evaluating the time-tagged SNSPD array with the TCSPC method, which however so far only enables this multiplexing technique for pulsed applications. Moreover, the maximum number of detectors in one chain is mainly limited by the repetition rate of the laser source. A concept to readout such a detector chain with the TCSPC method without the need of a pulsed laser would quite enforce application with continuous-wave light.

Up to now, the readout line was only connected on one side of the detector chain. The other end of the chain was shorted to the GND plane. In the SDD detector chain, the detector elements are embedded in a coplanar transmission line. By shorting the transmission line on one side a 1-port measurements is done as in the previous section and without shorting the transmission line a 2-port measurement is possible, as well. As reasoned in the last section, each detector element produces a positive pulse in the one direction of the readout line and in the same moment a negative pulse in the opposite direction. To measure the pulses on both sides of the detector chain, the readout (biasT, rf line, amplifier) is duplicated to enable a 2-port measurement. Depending on the local origin of the photon event in the detector chain, the arrival times varies depending on which output channel is used. Using as start and stop channel of the TCSPC electronics the two output lines of the SNSPD readout, two different measured times Δt are measured depending on the pulse releasing pixel. The two Δt have a delay time, which is in the range of $2 \cdot \tau_d$. The factor 2 comes from the fact that the time reference of the TCSPC measurement is always given from the opposite pixel pulse, which adds a further τ_d to the delay.

Fig. 5.39 shows such a double readout scheme. In Fig. 3.2 the dedicated transmission structure of the carrier board in the cryogenic dip stick can be seen, which includes two biasTs. The dip stick uses two independent rf lines to guide the rf signal out of the cryostat. On one rf line two cascaded ZFL-2500VH+ amplifiers are connected and on the other rf line two ZX60-33LN+.

Fig. 5.40 shows the oscilloscope measurements of this readout scheme. On both channels the typical pulse shapes are measured. The slight different in the shape of the pulse step comes from the different amplifier types used for readout. Setting the trigger of the oscilloscope on one of the pulse edges of the detector channel, the opposite pulse responses arrive delayed with $2 \cdot \tau_d$ (extended by the additional virtual delay ΔS), as expected.

The pulsed-laser reference signals in Fig. 5.40 are only recorded as reference of the delay times, but are not required for readout any more. The delay time between the readout channels can even be measured for cw excitation.

The measurements show that the step in the slope still exists due to the fact that at the position of the former GND short a complex impedance still exists, which leads to a mismatch at this position in the transmission chain. The dual measurements demonstrate that the step is just a characteristic of the superposed pulses and not a intrinsic detector characteristic. Comparing the two pulses of a single-shot measurement the step is only available on the response of one readout side.

The delay enlargement by factor 2 allows further reduction of the single delay lengths, which helps to keep L_{total} of the system smaller.

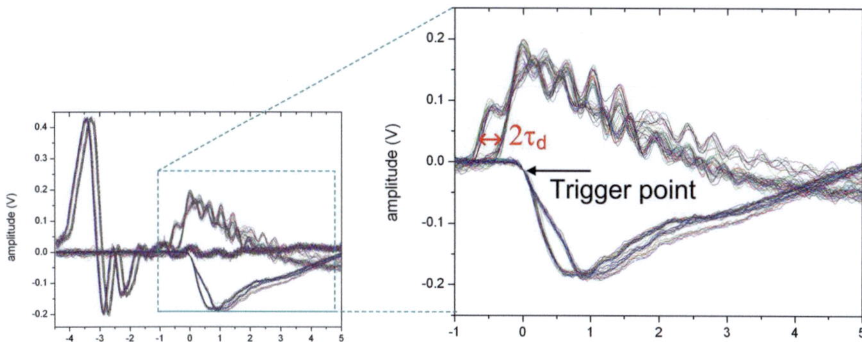

Figure 5.40: Oscilloscope measurement of a dual channel time-tagged experiment with two pixels. The time delay between the arriving pulses is twice the delay measured between the dedicated laser reference pulses.

5.4.9 Summary

A real-time readout of a multipixel SNSPD chip requires the acquisition of a counting event in the moment, when it occurs, and a continuous observation of the full detector. The cryogenic requirements of such a multipixel approach are the same as for all SNSPD readouts. In this subchapter a concept is introduced that bases on a chain of detector elements with delay lines in between. The delay lines add an additional time span to the arrival time of the detector response, which can be measured in a TCSPC measurement in relation to the reference pulse of a pulsed laser system. The delay lines are directly patterned in the superconducting thin film during the detector fabrication process. The feasibility of this readout concept is demonstrated with a 2-pixel and a 4-pixel chip. A deep analysis of the pulse shape showed that an rf model of the detector chain is required to explain the pulse shapes of each responding detector element. Typically, an SNSPD is readout by an rf line connected on one side of the detector meander. The time-tagged readout has shown that a detector produces on both sides of the detector meander a voltage pulse with inverted signs. Using a detector chain of several detector elements, these two pulses carry the delay time of the superconducting delay lines as relative time signature, which can be evaluated by the TCSPC method. The advancement of this measurement method is the removed necessity of a reference pulse of a pulsed laser excitation source. Consequently, the time-tagged multiplexing concept can be applied for cw radiation, as well.

6 Conclusion

Superconducting nanowire single-photon detectors (SNSPD) are meanwhile competitive candidates for single-photon detection applications. Especially, the tremendous effort in development has brought their detection efficiencies close to 100 % during the last years. In addition, SNSPDs gain from a very low dark count rate which reaches values $<< 1$ and they have very short dead times allowing high count rates of several hundreds of MHz and low jitter values down to 18 ps without afterpulsing effects .
The optimization of the detector performance was a step-wise improvement, which was deeply correlated with the increased understanding of the detection processes in the SNSPD meander. Physical understanding of superconducting and solid-state processes connected with engineering challenges were and still are required, to bring SNSPDs to the present performance opening further fields of applications in future.

The detector research required a high-qualitative cryogenic measurement equipment, for which reason a complete dip stick cryostat including operation electronics was developed, which is optimized for low-noise readout. With this setup it is possible to measure typical parameters like the detection efficiency, dark count rate, time jitter, and it enables special readout concepts required for the multipixel experiments.

For several years, the hot-spot model was the general explanation of the photon detection mechanism in an SNSPD. However, much detector research was only done on a single radiation wavelength. With the beginning of the broad spectral analysis, it was clear, that for the infrared regime the hot-spot model is not a suitable explanation. Consequently, theoretical descriptions of a vortex-assisted detection model were developed. The vortices, whose movement on the stripe is hampered by a potential barrier at the edge of the meander stripe, can nucleate and cross the stripe with a certain probability, which is enforced by the additional energy of an absorbed infrared photon. In this thesis, this infrared detection model is verified by a systematic analysis of the dependence of the detection efficiency on the film thickness and other operation parameters like the detector bias current. Furthermore, it was shown that the intrinsic detection efficiency, which is the detection efficiency of already absorbed photons, reaches up to 70 % for a 4 nm thick film, which means, that generally the low absorption of a typical detector design only limits the detection efficiency of an SNSPD.

Derived from the vortex detection mechanism, the idea has been developed by several groups that dark counts in an SNSPD originate from the same mechanism as the infrared detection, however, triggered by thermal fluctuations in the film. Based on the parameters of the vortex nucleation probability, several concepts were studied in this thesis to reduce the dark count rate. One concept focuses on the improved thermal coupling between SNSPD and the cooling bath to reduce the attempt rate of a vortex nucleation, which directly led to a reduction of the dark count rate. The next concept was the variation of the operation temperature. It has been shown that the change of temperature T, which defines the fluctuation energy $k_B \cdot T$ and the superconducting energy gap Δ, changes exponentially the probability of the dark counts. The final concept concentrated on the enlargement of the potential vortex barrier, by changing the square resistance R_S of the superconducting film. This was realized by a change of the film stoichiometry, which leads to an reduction of the dark count rate. With these concepts, the dark count rate has been decreased by several orders of magnitude for bias currents close to the critical current. At typical bias conditions of $0.95 \cdot I_C$, the dark count rate reaches values $\ll 1$ s^{-1}. All three concepts focused on an improvement of the dark count rate. Further results have demonstrated that the detection efficiency is not affected or even improves by this concepts. It can be concluded that all concepts together would provide a tremendous improvement of the SNSPD performance.

Understanding the detection mechanism and the most important performance parameters led to further challenges in developing of SNSPD applications: The feasibility of multipixel readout concepts of SNSPDs.
A multipixel readout can be generally defined in several application tasks. One task is the real-time framing, which means a measurement of the average count rate of a pixel matrix, which is evaluated in certain imaging frames, allowing video imaging. Another task is the evaluation of a multipixel matrix, or array in real time under continuous observation. Here, the single-photon information is measured and evaluated in the moment of the event and assigned to the corresponding pixel. The third approach, is the resolution of multi-photon events by a detector array/matrix, which would be quite interesting for astronomy. In this thesis, the focus is set on concepts for imaging and real-time single-photon readout.
During the last years, RSFQ became more and more interesting as suitable readout interface for SNSDs. RSFQ electronics is a superconducting electronics based on over-damped Josephson Junctions, which uses a single-flux quantum as data carrier. It was demonstrated for the first time that an RSFQ electronics is able to convert SNSPD signals into flux quantum data, which can be used for further data processing. In this thesis, an RSFQ pulse merger is introduced, which merges several detector channels to one output channel. Knowing that a continuous current biasing of the detector pixels removes the pixel assigning, a bias switching is added allowing time-gated measurements of several SNSPD detectors. Based on signal-theoretical concepts, a code division multiple access (CDMA) approach is adapted to the SNSPD readout. In an experiment the feasibility of such a concept is demonstrated with a 4-pixel matrix and the accuracy of the count rate evaluation is compared to

a time-division multiple access (TDMA). It has been shown that the CDMA approach improves the count rate evaluation of the strong counting pixels. For the weak-counting pixels, generally a worse resolution is expected compared to a TDMA, which however depends on the continuous flux of the optical source, since the CDMA approach has better averaging abilities. It can be summarized that RSFQ electronics is a suitable candidate for multipixel imaging applications. The task for the future will be to develop circuits with high junction densities allowing high pixel numbers or even an SNSPD integration in the RSFQ chip.

Real-time evaluation of single-photon events under continuous observation is not realizable with a classical RSFQ merger function, even a CDMA or TDMA approach can not solve this task. Moreover, for applications which only require small arrays of detectors, the RSFQ readout leads to a high system complexity, compared to the benefit of the multipixel ability. For this reason, a time-tagged multiplexing method has been developed and presented in this thesis, which evaluates an individual time delay signature in the picosecond range by using the time-correlated single-photon counting (TCSPC) method to assign a photon event in real-time to the corresponding pixel. The time-tag signature bases on delay lines between a chain of detector elements, which is patterned in the fabrication process of the detectors. The concept requires only one single bias line and one readout line for operation. The feasibility of this method was demonstrated with up to 4-pixels. A detailed pulse analysis was performed to explain the pulse shaping of the sequenced detector elements using a rf-circuit description. It has been found out that an SNSPD meander produces on both ends a pulse event with inverted signs. Using a detector chain embedded in a transmission line, it could be shown that the time-tag can be evaluated without the need of a reference pulse by applying the TCSPC method to the readout channels of the detector transmission line. Consequently, the method is suitable for applications with cw radiation, as well.

A BiasT based on superconducting planar structures

A requirement for all multipixel SNSPDs is an effective signal guiding on the detector chip to reach high pixel densities. The single-pixel design and the multipixel chips used in the previous sections have a coplanar waveguide for rf-signal transmission. This waveguide type is improper for high pixel densities due to the required GND plane on the top of the chip. For high pixel densities, a micro stripe design is more suitable. However, in any case there is a certain space required between the detector elements to enable the line routing between the SNSPD element and the dedicated readout line. A space-saving solution is the design of detector rows as in [43]. Several rows would lead to a matrix. In this case the bias, readout and GND lines can be guided parallel to the pixel rows. The local positioning of the detector elements should always be defined in connection with the final application, since to narrow located detector elements could lead to aliasing effects in the final image. According to section 3.2, a biasT is essential for each detector element to enable good signal transmission. A positioning of the indispensable biasT close to the detector element would reduce the routing complexity and the required space on the detector chip. Therefore, a biasT design integrated on the superconducting chip would help to optimize the multipixel chip design. The following work was done in co-work with Pavol Marco and is described in [133].

The standard operation concept of SNSPDs requires a separation of the biasing and the readout (Fig. A.1), which is realized by the biasT. The biasT e.g. from section 3.2 consists of an inductance L in form of a coil close to the detector element, which injects the bias current to the detector and simultaneously prevents the rf signal going along the bias path. Furthermore, a series capacitance C in the readout path realized by a soldered SMD component prevents a dc-current flow in the rf line. Inductance and capacitance are both components, which can be designed as lumped element from a planar film, as well. A lumped element is a component consisting of a geometric structure which is at least by a factor 20 smaller than the dedicated signal wavelength. Therefore, the planar structure behaves comparable to a hybrid component. The main advantage is that the planar components can be fabricated directly in the detector production process [133].
Since the following consideration covers the biasT design in general and not only with the focus of high pixel densities, the additional challenge of designing embedded biasTs in a coplanar transmission line are regarded, as well.

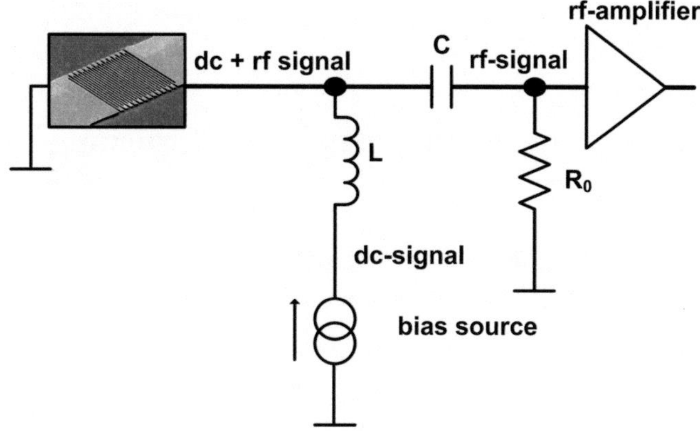

Figure A.1: Scheme of the readout with a biasT. The dc path is free of rf signals due to a low pass caused by an inductance L. The rf path is free of dc signals due to high pass caused by a capacitance C.

Planar capacity

Each of the biasT elements is considered separately.

The capacity C in the rf path of the biasT defines in connection with the line impedance R_0 a passive high-pass circuit. The 3 dB cutoff frequency of a first order high-pass is given by $f_{3dB} = \frac{1}{2\pi \cdot R_0 C}$. The resistance $R_0 = 50\ \Omega$ can be considered as the impedance of the transmission line. The response signals of typical SNSPD geometries require for good signal transmission a lower bandwidth limit of several MHz. This frequency range is similar to the lower limit of rf amplifiers which are available on the market for the lower GHz range. Therefore the biasT should be configured with a C leading to a lower bandwidth limit of $f_{3dB} = 50$ MHz. The dedicated capacitance C needs to be designed.

Since SNSPDs are a single layer device, the SNSPD fabrication process has only one film sputtering process. Maintaining the standard fabrication steps, only planar structures without any overlay, so called multi-layers, can be designed as capacitance.

The easiest design of a C is a small gap in the transmission line. A typical readout line width at IMS is 200 μm. The limitation of resolution of the contact photo lithography is 1 μm. The capacity C_{gap} can be simulated by the simulation tool "Sonnet em" to be about $C_{gap} = 43$ fF. With $R_0 = 50\ \Omega$ this would lead to a 3 dB high-pass frequency of $f_{3dB} = 74$ GHz, which is far too much to enable a transmission of a SNSPD pulse.

A shift of the f_{3dB} to lower frequencies can be realized by increasing the capacity C_{gap} because the R_0 is fix. One solution is to increase the length of the edge of the gap (which is equal to an enlarged area of a plate type capacitor) by designing finger capacities (inter-digital capacitor (IDC)) (see Fig. A.2). The limitations of a IDC are again the transmission line width and the limitations of the resolution of the contact photo lithography. The finger length is a crucial parameter for the definition of C_{IDC}. Theoretically, this length can be

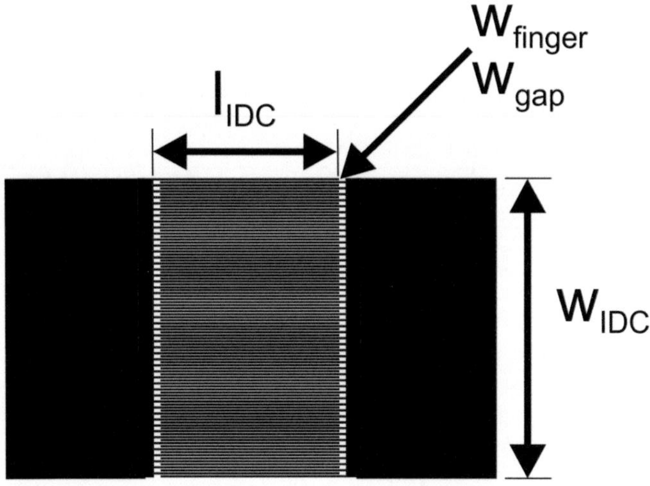

Figure A.2: Design of a planar finger capacity for simulation with Sonnet em. The capacity is integrated in the transmission line of a coplanar wave guide with a width of $w = 200$ μm and a gap of $g = 220$ μm. [133]

set quite long. However, the geometrical parameters of the capacitance should not exceed the geometrical dimensions of the other structures on the chip with regard to high pixel densities. The space consumption on a multipixel chip should not be dominated by the capacitance area.

A IDC capacitance C is simulated by Sonnet em with the parameters listed in Table A.1. The metal layer is a superconductor and therefore loss-less.

Table A.1: Parameters for the simulation of a planar IDC in a typical SNSPD coplanar line.

Parameter	Value
R_0	50 Ω
model of metal	Thick metal model: $d = 5$ nm
w_{IDC}	200 μm
l_{IDC}	135 μm
$w_{fingers}$	1 μm
w_{gap}	1 μm

The dedicated structure with fingers is shown in Fig. A.2 and covers an area of $A = 200$ μm x 135 μm. This simulation leads to a capacity $C \approx 600$ fF.

The simulated S-parameters are shown in Fig. A.3. The cutoff frequency of this structure is $f_{3dB} = 2.7$ GHz, which is still too high for SNSPD signal transmission. The simulated structure covers already an area, which is on the limit concerning compact high pixel readout. The capacitance is almost by factor 100 larger than the detector pixel geometry. Consequently, a one layer capacity is not suitable as biasT component as long as the patterning is

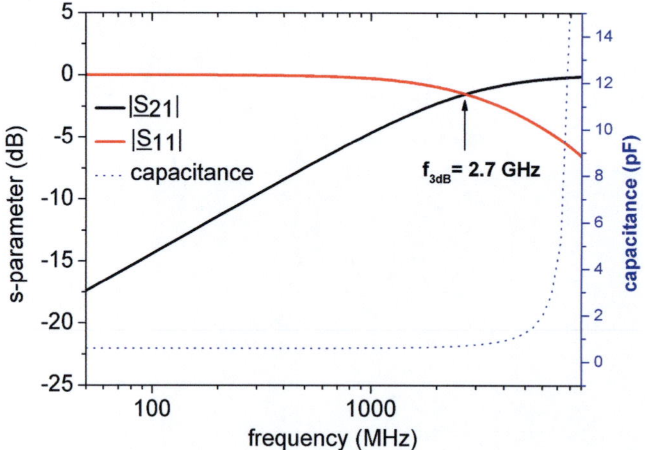

Figure A.3: Simulated capacity and S-parameter of the planar finger capacity in a coplanar waveguide. The transmission line has a cutoff frequency $f_{3dB} = 2.7$ GHz.

made by standard contact photo lithography and the chip size is limited. E-beam patterning can of course reach higher values because w_{finger} and w_{gap} can be reduced at the expense of longer patterning times for each device.

An more complex capacity concept consists of two planar layers with a very thin dielectric material in between. However, this design requires more fabrication steps including deposition of an isolating layer. A dielectric material for isolation could be AlN, SiO or SiO_2. The feasibility of such a structure is not analyzed here, since not all multipixel approaches urgently require a capacitance.

Planar inductance

The feasibility of a planar inductance is considered. The aim of the planar inductance is to enable the current injection to the detector without influencing the rf signal transmission. The planar inductance should be located close to the rf transmission line, which reduces the length of the branch line between transmission line and inductance. This prevents reflections back to the detector.

Estimation of the inductance dimension

The inductance defines a high-pass in connection with R_0. The f_{3dB} bandwidth is again defined by the SNSPD signal, which means a f_{3dB} of several MHz. Therefore, the aimed f_{3dB} cutoff is set to 40 MHz, which is similar to the value of the capacitance analysis.

The required inductance value has to be found out. In a first simulation, an inductance element is connected to a coplanar transmission line and shorted to GND. The grounding of the bias inductance is required to evaluate the transmission like a typical high-pass circuit.

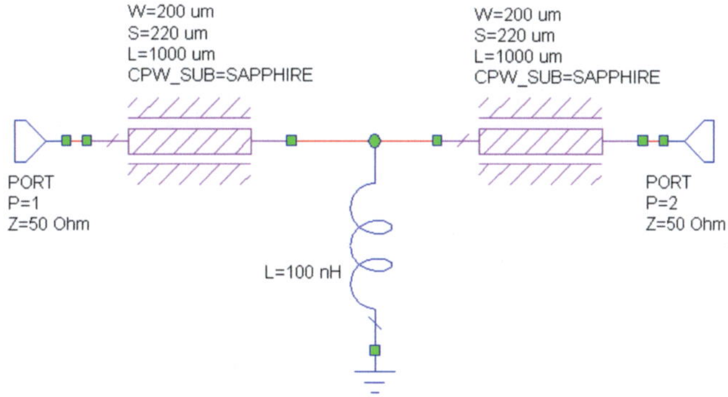

Figure A.4: Schematic for a simulation of the 50 Ω transmission line with a shorted inductance. [133]

A high damping of the transmission $|\underline{S}_{21}|$ means that the rf power vanishes trough the bias path. The ports of the transmission line are matched to $R_0 = 50\ \Omega$. For first estimation, a Microwave Office simulation is performed. The dedicated circuit is shown in A.4.

According to the simulation, an inductance of $L = 100$ nH is required for $f_{3dB} = 40$ MHz (see Fig. A.5). The simulation is based on the assumption of a two-side 50 Ω terminated transmission line. However, later one termination equates to the detector impedance itself, which has in the moment of a pulse event a dynamic impedance between zero and several hundreds of ohms, due to the dynamic of the normalconducting belt in the meander wire. Consequently, the simulated inductance can only be treated as an approximated value. Therefore, the planar inductance should have a certain reserve, which means a value of $L > 100$ nH. The next step is to develop a planar inductance geometry that accords with the required inductance L.

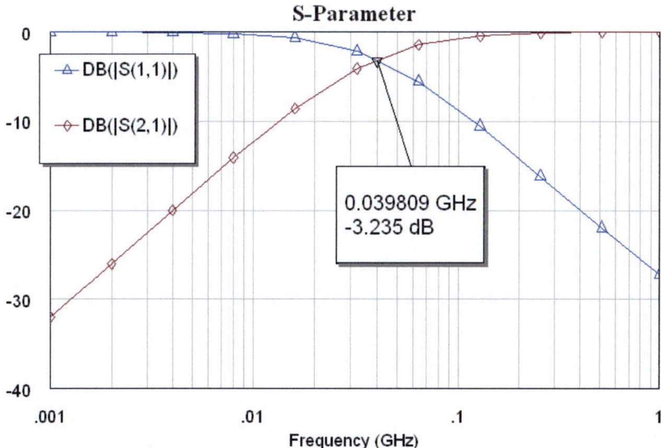

Figure A.5: Simulated $S_{21}(f)$ and $S_{11}(f)$ parameter of the transmission line with a 100 nH inductance as short. [133]

Design of a planar inductance

The inductance in superconductors consists of two parts: The geometric inductance L_{geo} and the kinetic inductance L_{kin} [130]. Common for both is the dependence on the length l_L, the width w_L and the thickness d of the structure: $L_{total} \propto \frac{l_L}{w_L \cdot d}$. Therefore, a high inductance can be reached designing long and narrow lines (the thickness d is the already given film thickness and therefore a fixed factor). The geometries are limited by the resolution of the photo lithography processes and by superconducting constraints, as well because the inductance is made from the same superconducting film as the detector structure. One can expect that the critical current density $j_{C,\,biasT}$ in the inductance is comparable (neglecting the proximity effect [134]) to $j_{C,\,SNSPD}$ of the detector meander. Based on this assumption, the width w_L of the inductance must be sufficient larger than the detector meander width w_{SNSPD}. The biasT inductance should stay superconducting in all operation modes and not behave as a detector or latch in the normalconducting state. For space-saving reasons, the line is meandered.

In [27], it is demonstrated that the kinetic inductance L_{kin} dominates the inductance of a SNSPD meander. From the measurements and geometries in [20], a suitable value for the square inductance can be derived to $L_\square = 120$ pH. A more exact estimation is not possible due to the unknown surface resistance R_S of the NbN used.

The number of squares for $L = 100$ nH can be calculated to:

$$n_{square} = \frac{L_L}{L_\square} = \frac{100 \text{ nH}}{120 \text{ pH}} = 833 = \frac{l_L}{w_L} \tag{A.1}$$

The photo lithographic minimum of resolution defines the width of the inductance to $w_L = 1$ µm and the required length of the meander is calculated to $l_L = 833 \cdot w_L = 833$ µm. One has to consider that a narrow meander structure reduces the inductance L_{geo} due to the superposition of the magnetic field of the parallel stripes with opposite current flow. Furthermore, there is a dependence of L_{kin} on the strength of current in the inductance [26].

To estimate the transmission behavior of the bias path with an inductive meander geometry (see Fig. A.6) a planar inductance is simulated with the simulation software Sonnet em. The inductance parameter are: $w_L : 1$ µm, $l_L : 1128$ µm. The line width is selected to get a value $L > 100$ nH. The surface of the metal layer is simulated to be loss-less with a square inductance $L_\square = 120$ pF. The inductance is embedded in a coplanar waveguide with $w = 35$ µm and $g = 20$ µm, which equals the embedding in the later GND plane.

The simulated S-parameters of the meander inductance are taken as sub-component in the transmission simulation of the biasT in Fig. A.4 instead of the lumped component model. The AWR simulation delivers in Fig. A.7 $|\underline{S}_{21}|$ and $|\underline{S}_{11}|$ of the rf path of the biasT. The cutoff of the high-pass of the inductance is shifted to 24 MHz compared to the simulation of Fig. A.5. The meander inductance extracted from the simulation data (see right axis of

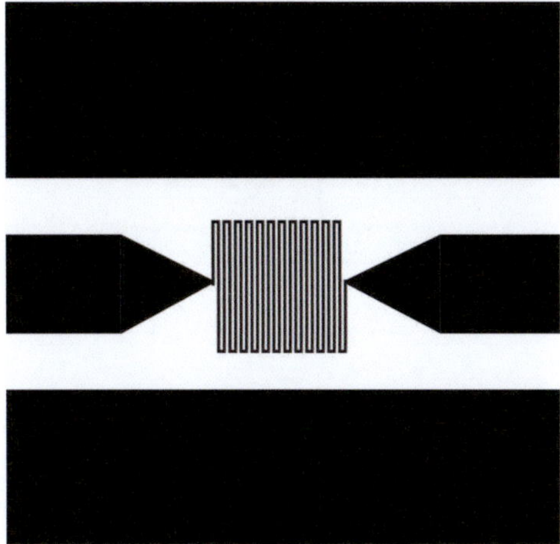

Figure A.6: Design of a planar meander with inductive characteristic. Line width:1 μm, line length: 1128 μm. Coplanar waveguide: width $w = 35$ μm, gap $g = 20$ μm. [133]

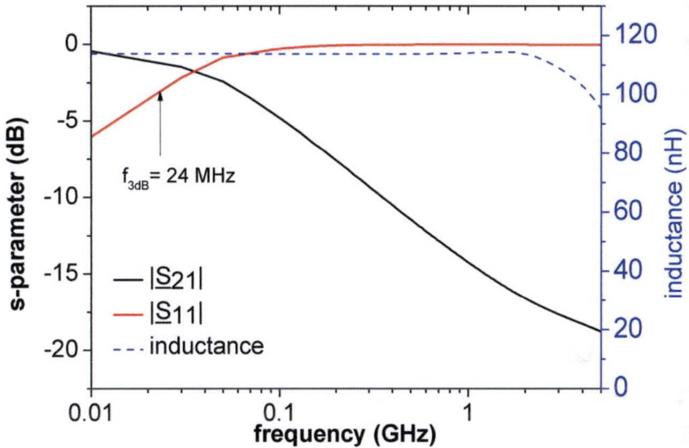

Figure A.7: Simulation of the transmission of the biasT using a planar meandered inductance (Fig. A.6). The designed meander leads to a cutoff frequency of 24 MHz and a inductance value of 115 nH.

Fig. A.5) delivers a value of around 115 nH, which is slightly larger than 100 nH as expected from the longer meander length l_L.

Fabrication of test structures of biasT

To evaluate a suitable geometric range around the geometric simulation parameters, a batch of samples covering a range of width and length parameters has to be be fabricated and tested under real conditions. This necessity is also enforced because the parameter $L_\square = 120$ pF is only an estimated value based on literature. An overview of the geometric parameters of fabricated samples is given in the Table A.2:

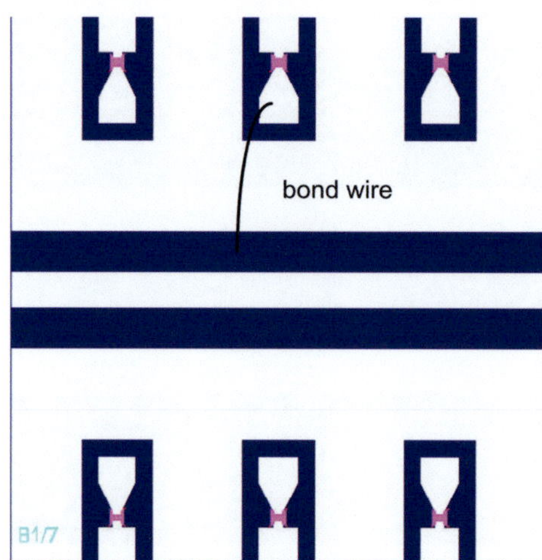

Figure A.8: Layout of a biasT test structure. It shows six inductance meanders with contact pads. The connection between inductance and transmission line will be realized by aluminum wire bond individually for each measurement. [133]

Table A.2: List of the parameters of the fabricated meanders as biasT inductance

length μm	width μm	number of squares
800	0.6	1333
800	0.8	1000
800	0.9	888
800	1.1	727
800	1.2	667
800	1.4	571

For the investigation of the variation of the meander width a chip of 3×3 mm^2 is designed with six inductance geometries on it. All inductances are separated from the transmission line, which is in the middle of the chip as shown in Fig. A.8. During measurement, always one inductance is bonded to the transmission line. Each inductance is measured in a separate cooling cycle.

Carrier box & measurement setup

The superconducting sample has to be embedded in a read out circuit inside a carrier box (CB), that allows a high frequency connection to the semi-rigid SMA cable of the dip stick cryostat. This connection is realized by an rf PCB which is mounted in a brass housing. The biasT sample is directly mounted on the brass holder by silver paste to guarantee good thermal contact. The rf PCB consists of a 50 Ω matched grounded coplanar line with dc-connection lines on one side of the coplanar line for the biasing path. The rf PCB design can be seen in Fig. A.9. The parameters of this plate are:

148

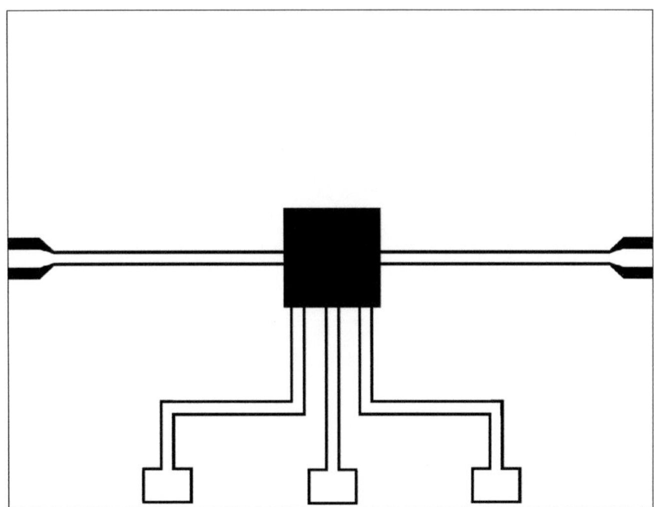

Figure A.9: Mask of the carrier board of the sample holder for characterization of the planar biasT inductance. [133]

Table A.3: Parameters of rf adapter plate

Parameter	Value
line impedance	50 Ω
substrate	Rogers TMM10i
ε_r	9.8
$tan(\delta)$	0.0025
thickness of substrate	635 μm
material	copper
thickness of wire	35 μm
width of line	220 μm
width of gap	150 μm

In the middle of the plate is a $3x3mm^2$ hole, which contains the biasT sample under investigation. The electrical contact between rf PCB and biasT sample is realized by indium bonds. The brass housing has two SMA connectors for rf transmission and dc connectors with an additional Π-low-pass-filter. During measurement the device can be covered by a cap to protect the sample. The cap has small holes as input for the liquid helium. The carrier box is shown in Fig. A.10.

Geometry and current dependence of biasT transmission
The samples are characterized by a 2-port rf measurement with an Agilent network analyzer. The principle of the measurement setup is demonstrated in Fig. A.11. The CB is connected to the two rf wires at the end of a cryogenic rf dip stick.

Before measurement, the network analyzer is calibrated including the cryogenic dip stick. The network analyzer consequently measures only the circuit between the connectors of the CB. To get a reference inside of the CB for de-embedding the biasT characteristic from parasitics like carrier board/sample transition, the pure transmission of the CB is once measured without bonded bias paths. The network analyzer applies on both ports a microwave power of -15dBm for all measurements in this section. The biasT is characterized in the same range as the typical SNSPD frequency range, which is selected to: 10 MHz to 5 GHz.

In Fig. A.12 the $|\underline{S}_{21}|$ parameter is analyzed. The width w of the meandered inductance is varied. The black line demonstrates the de-embedding reference, which is the pure transmission line without connected bias path. The course is almost flat except a parasitic dip at around 1.4 GHz. This parasitic consequently occurs in all measurements and can be neglected for evaluation. The reason for this dip probably lies in the not-matched interface between superconducting sample and carrier board or possible box resonances.
The dashed green line demonstrates a measurement of the transmission line where the bias line is directly bonded to the transmission line without an additional meander inductance. It is obvious that a direct bond connection of the bias path completely disturbs the system and transmits the energy into the wrong paths. Therefore, an additional inductance is essential for good rf transmission.
Considering the measurements of different meanders connected in the bias branch, $|\underline{S}_{21}|$ is flat and free of attenuation up to 3 GHz in the upper frequency range. The 3 dB cutoff frequency is outside of the measurement range, however, the start of the high-pass drop can be seen at around 40 MHz. A shift of the 3 dB cutoff frequency to lower frequencies with decreasing meander width (and therefore increasing inductance) can also be recognized.

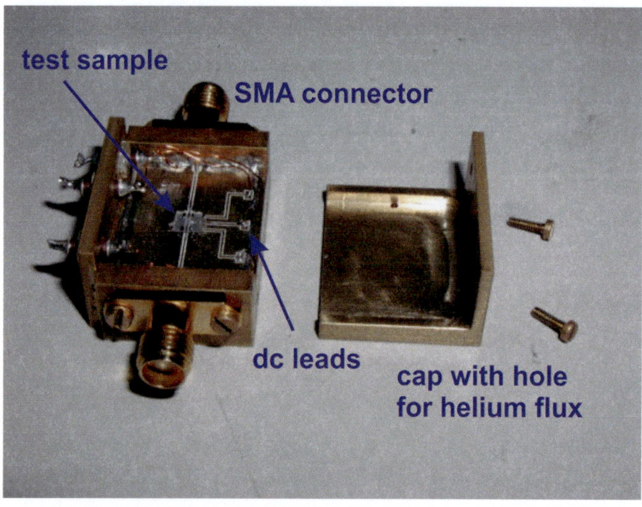

Figure A.10: BiasT mounted in a carrier box. [133]

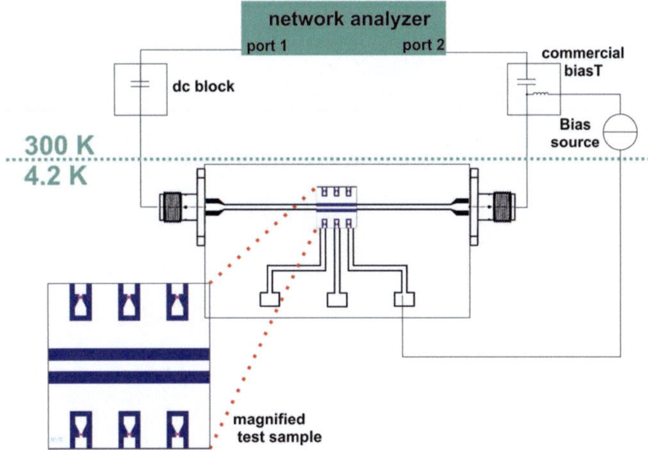

Figure A.11: Measurement setup for biasT rf characterization.

The covered bandwidth of the biasT is able to transmit the SNSPD signal. All inductance meanders deliver an adequate impedance to prevent the rf signal vanishing through the dc path in the required frequency range. The transmission is almost flat with slight drops in the range of 1 dB. In Fig. A.13 the analysis of $|\underline{S}_{11}|$ shows that the connection of the bias line without meander inductance produces a high reflecting behavior. However, the $|\underline{S}_{11}|$ parameters for the measurements with inductive meanders show a value of less than -10 dB over the complete frequency range, which is a suitable value for application.

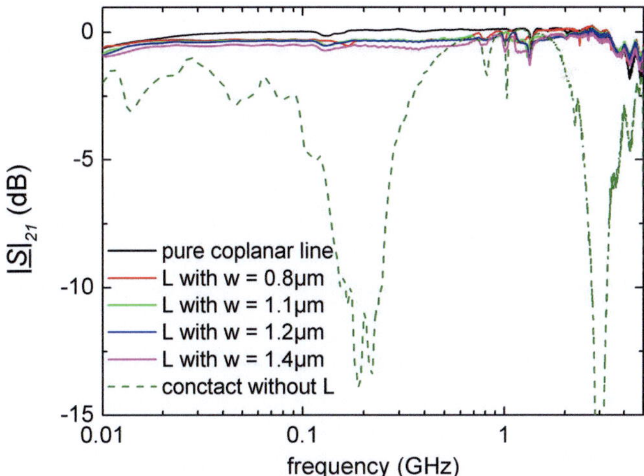

Figure A.12: S_{21} parameter measurement of the biasT. The black line is a reference line to de-embed parasitics, which are not caused by the biasT design. The green dashed line shows the behavior of the transmission, if the bias line is directly connected without meander inductance. All other curves presents a connected inductive meander with different width.

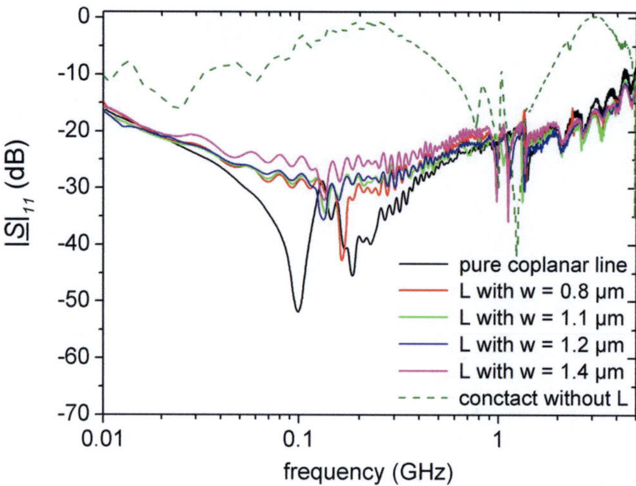

Figure A.13: S_{11} measurement of the biasT. Each curve presents a connected inductive meander with different width.

A change of the length l_L changes the number of $L_\square$. From geometric view, this is identical with a change of the width w_L. However, the width is the more critical parameter concerning superconducting constraints. One can conclude from the evaluation of the variation of w_L that the minimum width of $w_L = 1$ μm, which can be reached by standard conctact photo lithography in combination with a length $l_L = 800$ μm is sufficient for a suitable biasT inductance L.

Up to now, the dc path is only connected to a bias line, but not biased. Since the kinetic inductance L_{kin} is current dependent, it is helpful to analyze the dc current influence on the transmission behavior of the transmission line. The setup is modified with a dc block and a commercial biasT in the transmission path to realize a bias cycle with dc-blocked network analyzer ports. The extended setup can be seen in Fig. A.11. The additional components allow to inject a dc current on the transmission line going through the inductance of the planar biasT back to the bias source. The network analyzer is calibrated again before measurement.

The sample with the width $w = 1.0$ μm and the length $l = 800$ μm is used for the analysis. The bias current in the bias path is varied in the range of typical detector bias currents between 0 and 80 μA.

The results of the $|\underline{S}_{21}|$ are shown in Fig. A.14. The $|\underline{S}_{21}|$ transmission of the coplanar line improves with increasing bias current. That means, the value of the meander inductance is dependent on the current in the inductance, which is an attribute to L_{kin} [26]. Nevertheless, the transmission is for all currents close to 0 dB in the required frequency range.

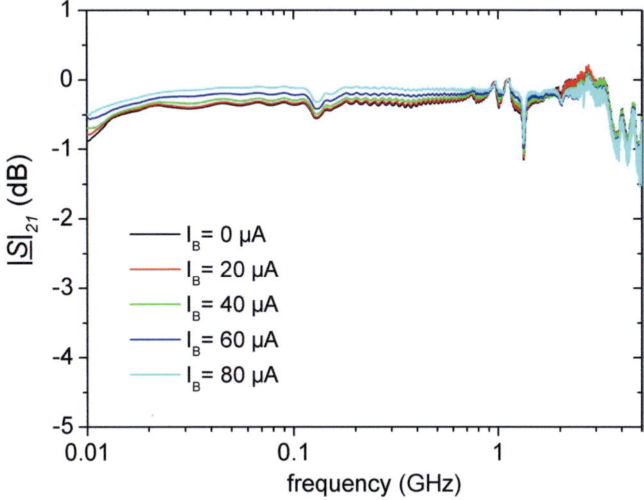

Figure A.14: Dependence of S_{21} on different bias currents in the inductive meander.

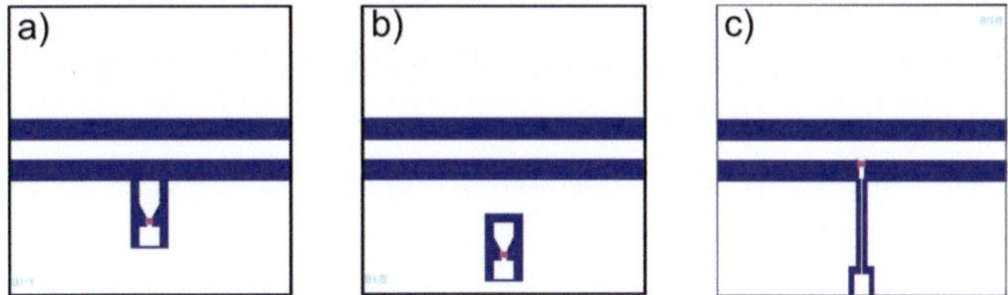

Figure A.15: E-line mask of a biasT design. a) and b): The bias inductance is bonded to the transmission line, the GND plane edge is continuous. c): The signal is guided trough the GND plane, the GND plane is interrupted at the edge of the gap. The interrupted GND edge can be continued by an air wire across the dc line. [133]

DC guiding in a microwave surrounding

Depending on the type of microwave transmission line, the routing of the dc line out of the chip is more or less complex. If the detector is embedded in a micro stripe or coupled stripe line, the inductance can directly be connected to the readout line of the detector element. In case of a coplanar waveguide, a GND plane surrounds the readout line, which needs to be crossed. To get a continuous GND current guiding along the edge of the gap of the coplanar line, either the inductive meander needs an air wire, or the GND layer has to be continued by an air wire, as long as a single-layer process is used for detector production.

One can consider three designs with different types of connections. In the design in Fig. A.15 a) and b), the inductive meander is separated from the transmission line. In one case, the GND-free island with the meander is directly placed at the edge of the GND plane to minimize the distance between transmission line and inductive meander. In the other case, the inductive meander has a distance of 250 μm from the gap edge. In Fig. A.15 c), the inductive meander is connected directly to the transmission line and guided through the GND plane. The GND current traveling along the edge of the gap can be optimized, if a bonded air wire is used to cross the dc line.

All variants are tested at cryogenic temperatures. A 2-port measurement is carried out again without applied bias current.

The $|\underline{S}_{21}|$ parameter of the transmission is shown in Fig. A.16. Again, the dip in the transmission at 1.4 GHz can be neglected because it exists also in the de-embedding reference measurement. All types of connections between inductive meander and transmission line have a good transmission. The maximum attenuation is less than 1 dB. The directly contacted inductive meander shows more flatness both with and without air wire in the GND path. The small loop way for the GND current around the inductance seems to be negligible. The inductive meander with bond connection to the transmission line shows more parasitic dips in the transmission, which is probably given by the reflection on the additional bond pad in front of the inductance and the bond wire.

One can conclude that it is possible to connect a bias branch either by interrupting the GND plane, or by using an air wire to connect the bias branch to the transmission line. The signal transmission is not negatively affected. A suitable selection is mainly defined by the number and density of pixels and the space of additional wire bonds on the final chip.

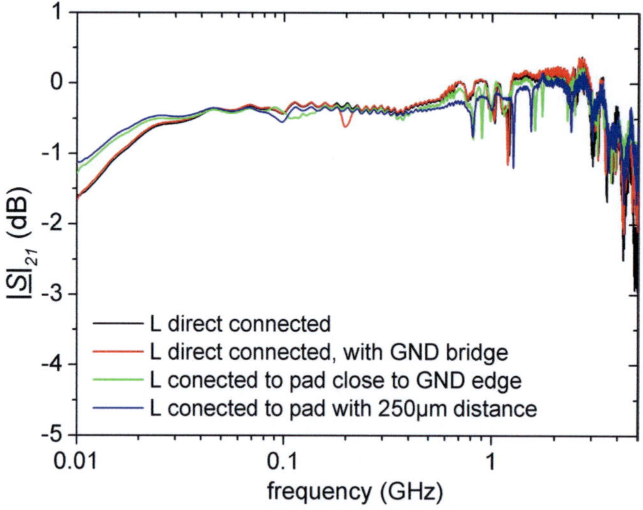

Figure A.16: S_{21} transmission of the different kinds of connection between readout line and inductive meander according to Fig. A.15.

Summary

An intelligent and space-saving design is a crucial requirement of a multipixel chip. In this chapter, a feasibility study of planar biasT components is analyzed. It is shown that a sufficient capacity made as single-layer planar structure is not realizable with sizes that allow high pixel numbers because the geometries of such a capacitance are more than factor 100 larger than the SNSPD structure itself. A multi-layer capacitance could solve this problem, however, at the expense of more complex fabrication.

A planar inductance is indeed possible to connect the bias line to the detector transmission line. Several geometries and connection methods are analyzed to find suitable design parameters. The fabrication process can be realized by standard contact photo lithography. It could be shown that a meander inductance with the dimension of 60×60 μm^2 is able to provide good transmission performance on the rf-transmission line to cover the frequency range required by an SNSPD. The size of such a structure is acceptable, even if the number of pixels on a chip increases. The meander dimensions would allow up to 60 pixel on one edge of a $3x3mm^2$ sample chip. A staggered order at the edge of the chip can even increase this number. Consequently, it is possible to design complete multipixel solutions on the superconducting film, including a bias wiring which only requires wire bond connections on the edge of the chip to connect them to the bias supply lines.

B List of Figures

C List of Tables

D Bibliography

[1] R. Bauval and A. Gilbert. *The Orion Mystery: Unlocking the Secrets of the Pyramids.* Arrow, 1995.

[2] C. Krafft, B. Dietzek, and J. Popp. "Raman and CARS microspectroscopy of cells and tissues". *Analyst*, **134**, 1046–1057, 2009.

[3] S. Yang, L. You, M. Zhang, and J. Wang. "Research of Single Photon Detectors Applied in Quantum Communication". In *Informatics and Management Science IV*, volume 207 of *Lecture Notes in Electrical Engineering*, pp. 19–27. Springer London, 2013.

[4] M. Ghioni, A. Gulinatti, I. Rech, F. Zappa, and S. Cova. "Progress in Silicon Single-Photon Avalanche Diodes". *Selected Topics in Quantum Electronics, IEEE Journal of*, **13**, 852–862, 2007.

[5] http://www.vertilon.com, ".

[6] H. Kamerlingh-Onnes. "Comm. Leiden 124c". *Comm. Leiden*, **124c**, 2011.

[7] B. D. Josephson. "Possible new effects in superconductive tunnelling". *Physics Letters*, **1**, 251–253, July 1962.

[8] J. Bednorz and K. Müller. "Possible high T_c superconductivity in the Ba-La-Cu-O system". *Zeitschrift für Physik B Condensed Matter*, **64**, 189–193, 1986.

[9] U. Pracht, E. Heintze, C. Clauss, D. Hafner, R. Bek, D. Werner, S. Gelhorn, M. Scheffler, M. Dressel, D. Sherman, B. Gorshunov, K. Il'in, D. Henrich, and M. Siegel. "Electrodynamics of the Superconducting State in Ultra-Thin Films at THz Frequencies". *Terahertz Science and Technology, IEEE Transactions on*, **3**, 269–280, 2013.

[10] A. D. Semenov, G. N. Gol'tsman, and R. Sobolewski. "Hot-electron effect in superconductors and its applications for radiation sensors". *Superconductor Science and Technology*, **15**, R1, 2002.

[11] A. D. Semenov, R. S. Nebosis, Y. P. Gousev, M. A. Heusinger, and K. F. Renk. "Analysis of the nonequilibrium photoresponse of superconducting films to pulsed radiation by use of a two-temperature model". *Phys. Rev. B*, **52**, 581–590, Jul 1995.

[12] C. Natarajan, M. Tanner, and R. Hadfield. "Superconducting nanowire single-photon detectors: physics and applications". *Superconductor Science and Technology*, **25**, 063001, 2012.

[13] K. Inderbitzin, A. Engel, and A. Schilling. "Soft X-Ray Single-Photon Detection With Superconducting Tantalum Nitride and Niobium Nanowire". *arXiv:1210.3675*, October 2012.

[14] G. Gol'tsman, O. Minaeva, A. Korneev, M. Tarkhov, I. Rubtsova, A. Divochiy, I. Milostnaya, G. Chulkova, N. Kaurova, B. Voronov, D. Pan, J. Kitaygorsky, A. Cross, A. Pearlman, I. Komissarov, W. Slysz, M. Wegrzecki, P. Grabiec, and R. Sobolewski. "Middle-Infrared to Visible-Light Ultrafast Superconducting Single-Photon Detectors". *Applied Superconductivity, IEEE Transactions on*, **17**, 246 –251, june 2007.

[15] K. Suzuki, S. Miki, S. Shiki, Y. Kobayashi, K. Chiba, Z. Wang, and M. Ohkubo. "Ultrafast ion detection by superconducting NbN thin-film nanowire detectors for time-of-flight mass spectrometry". *Physica C: Superconductivity*, **468**, 2001 – 2003, 2008.

[16] W. Pernice, C. Schuck, O. Minaeva, M. Li, G. Goltsman, A. Sergienko, and H. Tang. "High-speed and high-efficiency travelling wave single-photon detectors embedded in nanophotonic circuits". *Nat Commun*, **3**, December 2012.

[17] M. J. Stevens, R. H. Hadfield, R. Schwall, J. Gupta, S. W. Nam, and R. Mirin. "Fast lifetime measurements of infrared emitters with low-jitter superconducting single photon detectors". In *Lasers and Electro-Optics, 2006 and 2006 Quantum Electronics and Laser Science Conference. CLEO/QELS 2006. Conference on*, pp. 1–2, 2006.

[18] M. Tinkham. *Introduction to superconductivity*. McGraw-Hill, NY, 1996.

[19] A. Semenov, G. Goltsman, and A. Korneev. "Quantum detection by current carrying superconducting film". *Physica C: Superconductivity*, **351**, 349 – 356, 2001.

[20] A. Semenov, P. Haas, H.-W. Hübers, K. Ilin, M. Siegel, A. Kirste, D. Drung, T. Schurig, and A. Engel. "Intrinsic quantum efficiency and electro-thermal model of a superconducting nanowire single-photon detector". *Journal of Modern Optics*, **56**, 345–351, 2009.

[21] A. Semenov, A. Engel, H.-W. Hübers, K. Il'in, and M. Siegel. "Probability of the resistive state formation caused by absorption of a single-photon in current-carrying superconducting nano-strips". *European Physical Journal B*, **47**, 495, 2005.

[22] H. L. Hortensius, E. F. C. Driessen, T. M. Klapwijk, K. K. Berggren, and J. R. Clem. "Critical-current reduction in thin superconducting wires due to current crowding". *Applied Physics Letters*, **100**, 182602, 2012.

[23] B. Baek, A. E. Lita, V. Verma, and S. W. Nam. "Superconducting a-W_xSi_{1-x} nanowire single-photon detector with saturated internal quantum efficiency from visible to 1850 nm". *Applied Physics Letters*, **98**, 251105, 2011.

[24] M. Ejrnaes, A. Casaburi, R. Cristiano, O. Quaranta, S. Marchetti, N. Martucciello, S. Pagano, A. Gaggero, F. Mattioli, R. Leoni, P. Cavalier, and J. C. Villégier. "Timing jitter of cascade switch superconducting nanowire single photon detectors". *Applied Physics Letters*, **95**, 132503, 2009.

[25] A. K. Bhatnagar and E. A. Stern. "Experimental Investigation of Magnetic Field Dependence of Supercurrent in Multiply Connected Superconducting Films". *Phys. Rev. B*, **8**, 1061–1079, Aug 1973.

[26] J. R. Clem and V. G. Kogan. "Kinetic impedance and depairing in thin and narrow superconducting films". *Phys. Rev. B*, **86**, Nov 2012.

[27] A. J. Kerman, E. A. Dauler, W. E. Keicher, J. K. W. Yang, K. K. Berggren, G. Gol'tsman, and B. Voronov. "Kinetic-inductance-limited reset time of superconducting nanowire photon counters". *Applied Physics Letters*, **88**, 111116, 2006.

[28] J.-J. Chang and D. J. Scalapino. "Kinetic-equation approach to nonequilibrium superconductivity". *Phys. Rev. B*, **15**, 2651–2670, Mar 1977.

[29] K. S. Il'in, M. Lindgren, M. Currie, A. D. Semenov, G. N. Gol'tsman, R. Sobolewski, S. I. Cherednichenko, and E. M. Gershenzon. "Picosecond hot-electron energy relaxation in NbN superconducting photodetectors". *Applied Physics Letters*, **76**, 2752–2754, 2000.

[30] R. W. Heeres and V. Zwiller. "Superconducting detector dynamics studied by quantum pump-probe spectroscopy". *Applied Physics Letters*, **101**, 112603, 2012.

[31] S. B. Kaplan. "Acoustic matching of superconducting films to substrates". *Journal of Low Temperature Physics*, **37**, 343–365, 1979.

[32] N. Perrin and C. Vanneste. "Response of superconducting films to a periodic optical irradiation". *Phys. Rev. B*, **28**, 5150–5159, Nov 1983.

[33] M. Danerud, D. Winkler, M. Lindgren, M. Zorin, V. Trifonov, B. S. Karasik, G. N. Gol'tsman, and E. M. Gershenzon. "Nonequilibrium and bolometric photoresponse in patterned $YBa_2Cu_3O_{7-\delta}$ thin films". *Journal of Applied Physics*, **76**, 1902–1909, 1994.

[34] K. S. Il'in, I. I. Milostnaya, A. A. Verevkin, G. N. Gol'tsman, E. M. Gershenzon, and R. Sobolewski. "Ultimate quantum efficiency of a superconducting hot-electron photodetector". *Applied Physics Letters*, **73**, 3938–3940, 1998.

[35] J. Yang, A. Kerman, E. Dauler, V. Anant, K. Rosfjord, and K. Berggren. "Modeling the Electrical and Thermal Response of Superconducting Nanowire Single-Photon Detectors". *Applied Superconductivity, IEEE Transactions on*, **17**, 581 –585, june 2007.

[36] F. Pedrotti, L. Pedrotti, W. Bausch, and H. Schmidt. *Optik für Ingenieure*. Springer-Verlag, 2008.

[37] V. Anant, A. J. Kerman, E. A. Dauler, J. K. W. Yang, K. M. Rosfjord, and K. K. Berggren. "Optical properties of superconducting nanowire single-photon detectors". *Opt. Express*, **16**, 10750–10761, Jul 2008.

[38] A. Semenov, P. Haas, B. Günther, H.-W. Hübers, K. Il'in, M. Siegel, A. Kirste, J. Beyer, D. Drung, T. Schurig, and A. Smirnov. "An energy-resolving superconducting nanowire photon counter". *Superconductor Science and Technology*, **20**, 919, 2007.

[39] A. Semenov, B. Günther, U. Böttger, H.-W. Hübers, H. Bartolf, A. Engel, A. Schilling, K. Ilin, M. Siegel, R. Schneider, D. Gerthsen, and N. A. Gippius. "Optical and transport properties of ultrathin NbN films and nanostructures". *Phys. Rev. B*, **80**, 054510, Aug 2009.

[40] A. Christ, T. Zentgraf, J. Kuhl, S. G. Tikhodeev, N. A. Gippius, and H. Giessen. "Optical properties of planar metallic photonic crystal structures: Experiment and theory". *Phys. Rev. B*, **70**, 125113, Sep 2004.

[41] B. Baek, J. A. Stern, and S. W. Nam. "Superconducting nanowire single-photon detector in an optical cavity for front-side illumination". *Applied Physics Letters*, **95**, 191110, 2009.

[42] K. M. Rosfjord, J. K. W. Yang, E. A. Dauler, A. J. Kerman, V. Anant, B. M. Voronov, G. N. Gol'tsman, and K. K. Berggren. "Nanowire single-photon detector with an integrated optical cavity and anti-reflection coating". *Opt. Express*, **14**, 527–534, Jan 2006.

[43] E. Dauler, B. Robinson, A. Kerman, J. Yang, E. Rosfjord, V. Anant, B. Voronov, G. Gol'tsman, and K. Berggren. "Multi-Element Superconducting Nanowire Single-Photon Detector". *Applied Superconductivity, IEEE Transactions on*, **17**, 279 –284, june 2007.

[44] H. Terai, T. Yamashita, S. Miki, K. Makise, and Z. Wang. "Low-jitter single flux quantum signal readout from superconducting single photon detector". *Opt. Express*, **20**, 20115–20123, Aug 2012.

[45] J. Kitaygorsky, I. Komissarov, A. Jukna, D. Pan, O. Minaeva, N. Kaurova, A. Divochiy, A. Korneev, M. Tarkhov, B. Voronov, I. Milostnaya, G. Gol'tsman, and R. Sobolewski. "Dark Counts in Nanostructured NbN Superconducting Single-Photon Detectors and Bridges". *Applied Superconductivity, IEEE Transactions on*, **17**, 275 –278, june 2007.

[46] A. Engel, A. Semenov, H. W. Hübers, K. Il'in, and M. Siegel. "Superconducting single-photon detector for the visible and infrared spectral range". *Journal of Modern Optics*, **51**, 1459–1466, 2004.

[47] H. Bartolf, A. Engel, A. Schilling, K. Il'in, M. Siegel, H.-W. Hübers, and A. Semenov. "Current-assisted thermally activated flux liberation in ultrathin nanopatterned NbN superconducting meander structures". *Phys. Rev. B*, **81**, 024502, Jan 2010.

[48] A. Engel, A. Semenov, H.-W. Hübers, K. Il'in, and M. Siegel. "Fluctuation effects in superconducting nanostrips". *Physica C: Superconductivity*, **444**, 12 – 18, 2006.

[49] M. K. Akhlaghi, H. Atikian, A. Eftekharian, M. Loncar, and A. H. Majedi. "Reduced dark counts in optimized geometries for superconducting nanowire single photon detectors". *Opt. Express*, **20**, 23610–23616, Oct 2012.

[50] W. Buckel and R. Kleiner. *Supraleitung: Grundlagen und Anwendungen. 6. vollst. überarb. u. erw. Auflage*. WILEY-VCH Verlag GmbH & Co. KGaA, Weinheim, 2004.

[51] A. J. Kerman, J. K. W. Yang, R. J. Molnar, E. A. Dauler, and K. K. Berggren. "Electrothermal feedback in superconducting nanowire single-photon detectors". *Phys. Rev. B*, **79**, 100509, Mar 2009.

[52] A. Semenov, P. Haas, K. Ilin, H.-W. Hübers, M. Siegel, A. Engel, and A. Smirnov. "Energy resolution and sensitivity of a superconducting quantum detector". *Physica C: Superconductivity*, **460-462, Part 2**, 1491 – 1492, 2007.

[53] A. Pearlman, A. Cross, W. Slysz, J. Zhang, A. Verevkin, M. Currie, A. Korneev, P. Kouminov, K. Smirnov, B. Voronov, G. Gol'tsman, and R. Sobolewski. "Gigahertz counting rates of NbN single-photon detectors for quantum communications". *Applied Superconductivity, IEEE Transactions on*, **15**, 579 – 582, june 2005.

[54] E. A. Dauler, A. J. Kerman, B. S. Robinson, J. K. Yang, B. Voronov, G. Goltsman, S. A. Hamilton, and K. K. Berggren. "Photon-number-resolution with sub-30-ps timing using multi-element superconducting nanowire single photon detectors". *Journal of Modern Optics*, **56**, 364–373, 2009.

[55] J. A. O'Connor, M. G. Tanner, C. M. Natarajan, G. S. Buller, R. J. Warburton, S. Miki, Z. Wang, S. W. Nam, and R. H. Hadfield. "Spatial dependence of output pulse delay in a niobium nitride nanowire superconducting single-photon detector". *Applied Physics Letters*, **98**, 201116, 2011.

[56] J. R. Clem and K. K. Berggren. "Geometry-dependent critical currents in superconducting nanocircuits". *Phys. Rev. B*, **84**, 174510, Nov 2011.

[57] W. Becker, A. Bergmann, H. Wabnitz, D. Grosenick, and A. Liebert. "High count rate multichannel TCSPC for optical tomography". In *Biomedical Topical Meeting*, p. PD9. Optical Society of America, 2002.

[58] S. Jahanmirinejad, G. Frucci, F. Mattioli, D. Sahin, A. Gaggero, R. Leoni, and A. Fiore. "Photon-number resolving detector based on a series array of superconducting nanowires". *Applied Physics Letters*, **101**, 072602, 2012.

[59] H. Terai, S. Miki, and Z. Wang. "Readout Electronics Using Single-Flux-Quantum Circuit Technology for Superconducting Single-Photon Detector Array". *Applied Superconductivity, IEEE Transactions on*, **19**, 350 –353, june 2009.

[60] Hirotaka Terai and Shigehito Miki and Taro Yamashita and Kazumasa Makise and Zhen Wang. "Demonstration of single-flux-quantum readout operation for superconducting single-photon detectors". *Applied Physics Letters*, **97**, 112510, 2010.

[61] C. Enss and S. Hunklinger. *Low-Temperaure Physics*. Springer Berlin / Heidelberg New York, 2005.

[62] J. Clarke and A. I. Braginski. *The SQUID Handbook Vol.I Fundamentals and Technology of SQUIDs and SQUID Systems*. WILEY-VCH, 2004.

[63] S. Wuensch, E. Crocoll, M. Schubert, G. Wende, H.-G. Meyer, and M. Siegel. "Design and development of a cryogenic semiconductor amplifier for interfacing RSFQ circuits at 4.2 K". *Superconductor Science and Technology*, **20**, S356, 2007.

[64] T. J. Seebeck. "Magnetische Polarisation der Metalle und Erze durch Temperatur-Differenz.". *Abhandlungen der Preußischen Akademie der Wissenschaften zu Berlin*, 1823.

[65] M. Arndt. "Entwurf und Aufbau einer rauscharmen Bias-Stromquelle für supraleitende Detektoren". Master's thesis, KIT, 2010.

[66] N. R. Gemmell, A. McCarthy, B. Liu, M. G. Tanner, S. D. Dorenbos, V. Zwiller,
M. S. Patterson, G. S. Buller, B. C. Wilson, and R. H. Hadfield. "Singlet oxygen
luminescence detection with a fiber-coupled superconducting nanowire single-photon
detector". *Opt. Express*, **21**, 5005–5013, Feb 2013.

[67] X. Hu, T. Zhong, J. E. White, E. A. Dauler, F. Najafi, C. H. Herder, F. N. C. Wong,
and K. K. Berggren. "Fiber-coupled nanowire photon counter at 1550 nm with 24%
system detection efficiency". *Opt. Lett.*, **34**, 3607–3609, Dec 2009.

[68] A. J. Miller, A. E. Lita, B. Calkins, I. Vayshenker, S. M. Gruber, and S. W. Nam.
"Compact cryogenic self-aligning fiber-to-detector coupling with losses below one
percent". *Opt. Express*, **19**, 9102–9110, May 2011.

[69] G. Bachar, I. Baskin, O. Shtempeluck, and E. Buks. "On-fiber superconducting
nanowire single photon detector". In *2012 Applied Superconductivity Conference,
Portland, Oregon*, 2012.

[70] www.thorlabs.de/navigation.cfm?guide_id=26, ".

[71] D. Henrich. *Material and Geometry Dependencies of Superconducting Nanowire
Single-Photon Detector Performance*. PhD thesis, KIT, 2013.

[72] S. Dörner. "Erhöhung der spektralen Bandbreite von NbN SSPDs". Master's thesis,
KIT, 2012.

[73] W. Becker, A. Bergmann, M. Hink, K. König, K. Benndorf, and C. Biskupp. "Fluorescence Lifetime Imaging by Time-Correlated Single-Photon Counting". *Microscopy Research and Technique*, **63**, 58–66, 2004.

[74] http://www.becker hickl.de, ".

[75] Q. Zhao, L. Zhang, T. Jia, L. Kang, W. Xu, J. Chen, and P. Wu. "Intrinsic timing
jitter of superconducting nanowire single-photon detectors". *Applied Physics B*, **104**,
673–678, 2011.

[76] G. Bertolini and A. Coche. *Semiconductor Detectors*. North-Holland, Amsterdam,
1968.

[77] D. Gedcke and W. McDonald. "A constant fraction of pulse height trigger for opti-
mum time resolution". *Nuclear Instruments and Methods*, **55**, 377 – 380, 1967.

[78] W. Becker. "The bh TCSPC Handbook", 2006.

[79] M. Arndt. "Entwicklung eines Mischsignal-Messmoduls für Jittermessungen im Picosekundenbereich". Master's thesis, KIT, 2011.

[80] G. Deutscher. "Andreev–Saint-James reflections: A probe of cuprate superconductors". *Rev. Mod. Phys.*, **77**, 109–135, Mar 2005.

[81] R. Romestain, B. Delaet, P. Renaud-Goud, I. Wang, C. Jorel, J.-C. Villegier, and J.-P. Poizat. "Fabrication of a superconducting niobium nitride hot electron bolometer for single-photon counting". *New Journal of Physics*, **6**, 129, 2004.

[82] J. Zhang, W. Słysz, A. Pearlman, A. Verevkin, R. Sobolewski, O. Okunev, G. Chulkova, and G. N. Gol'tsman. "Time delay of resistive-state formation in superconducting stripes excited by single optical photons". *Phys. Rev. B*, **67**, 132508, Apr 2003.

[83] S. Miki, M. Fujiwara, M. Sasaki, B. Baek, A. J. Miller, R. H. Hadfield, S. W. Nam, and Z. Wang. "Large sensitive-area NbN nanowire superconducting single-photon detectors fabricated on single-crystal MgO substrates". *Applied Physics Letters*, **92**, 061116, 2008.

[84] M. G. Tanner, C. M. Natarajan, V. K. Pottapenjara, J. A. O'Connor, R. J. Warburton, R. H. Hadfield, B. Baek, S. Nam, S. N. Dorenbos, E. B. Ureña, T. Zijlstra, T. M. Klapwijk, and V. Zwiller. "Enhanced telecom wavelength single-photon detection with NbTiN superconducting nanowires on oxidized silicon". *Applied Physics Letters*, **96**, 221109, 2010.

[85] S. A. Wolf, F. J. Rachford, and M. Nisenoff. "Niobium nitride thin-film SQUID's biased at 20 MHz and 9.2 GHz". *Journal of Vacuum Science and Technology*, **15**, 386–388, 1978.

[86] S. Park and T. Geballe. "T_c depression in thin Nb films". *Physica B+C*, **135**, 108–112, 185.

[87] L. N. Cooper. "Superconductivity in the Neighborhood of Metallic Contacts". *Phys. Rev. Lett.*, **6**, 689–690, Jun 1961.

[88] Y. V. Fominov and M. V. Feigel'man. "Superconductive properties of thin dirty superconductor–normal-metal bilayers". *Phys. Rev. B*, **63**, 094518, Feb 2001.

[89] K. Ilin, R. Schneider, D. Gerthsen, A. Engel, H. Bartolf, A. Schilling, A. Semenov, H.-W. Huebers, B. Freitag, and M. Siegel. "Ultra-thin NbN films on Si: crystalline and superconducting properties". *Journal of Physics: Conference Series*, **97**, 012045, 2008.

[90] K. Ilin, M. Siegel, A. Semenov, A. Engel, and H.-W. Hübers. "Suppression of super-conductivity in Nb and NbN thin-film nano-structures". *Inst. Phys. Conf. Ser.*, **181**, 2895, 2004.

[91] J. Bardeen. "Critical Fields and Currents in Superconductors". *Rev. Mod. Phys.*, **34**, 667–681, Oct 1962.

[92] R. Romestain, B. Delaet, P. Renaud-Goud, I. Wang, C. Jorel, J.-C. Villegier, and J.-P. Poizat. "Fabrication of a superconducting niobium nitride hot electron bolometer for single-photon counting". *New Journal of Physics*, **6**, 129, 2004.

[93] D. Roditchev. "private communications of A. Semenov (2009). In the NbN films used in this study, the energy gap with the mean value $\Delta(0) = 2.1 k_B \cdot T_C$ has been found to vary within 15% on the lateral scale larger than 20 nm. The results will be published elsewhere.".

[94] M. Y. Kupriyanov and V. F. Lukichev. "Temperature dependence of pair-breaking current in superconductors". *Sov. J. Low Temp. Phys.*, **6(4)**, 210–14, 1980.

[95] A. J. Kerman, E. A. Dauler, J. K. W. Yang, K. M. Rosfjord, V. Anant, K. K. Berggren, G. N. Gol'tsman, and B. M. Voronov. "Constriction-limited detection efficiency of superconducting nanowire single-photon detectors". *Applied Physics Letters*, **90**, 101110, 2007.

[96] A. D. Semenov, P. Haas, H.-W. Hübers, K. S. Ilin, M. Siegel, A. Kirste, T. Schurig, and R. Herrmann. "Improved energy resolution of a superconducting single-photon detector". volume 7021, p. 70211E. SPIE, 2008.

[97] F. Tafuri, J. Kirtley, D. Born, D. Stornaiuolo, P. Medaglia, P. Orgiani, G. Balestrino, and V. Kogan. "Dissipation in ultra-thin current-carrying superconducting bridges; evidence for quantum tunneling of Pearl vortices". *EPL (Europhysics Letters)*, **73**, 948, 2006.

[98] A. Engel, H. Bartolf, A. Schilling, A. Semenov, H.-W. Hübers, K. Il'in, and M. Siegel. "Magnetic vortices in superconducting photon detectors". *Journal of Modern Optics*, **56**, 352–357, 2009.

[99] K. K. Likharev. "Superconducting weak links". *Rev. Mod. Phys.*, **51**, 101–159, Jan 1979.

[100] C. Qiu and T. Qian. "Numerical study of the phase slip in two-dimensional super-conducting strips". *Phys. Rev. B*, **77**, 174517, May 2008.

[101] J. Pearl. "Current distribution in superconducting films carrying quantized fluxoids". *Applied Physics Letters*, **5**, 65–66, 1964.

[102] K. K. Likharev. ". *Sov. Radiophys.*, **14**, 722, 1972.

[103] G. Stan, S. B. Field, and J. M. Martinis. "Critical Field for Complete Vortex Expulsion from Narrow Superconducting Strips". *Phys. Rev. Lett.*, **92**, 097003, Mar 2004.

[104] A. D. Semenov, P. Haas, H.-W. Hübers, K. Ilin, M. Siegel, A. Kirste, T. Schurig, and A. Engel. "Vortex-based single-photon response in nanostructured superconducting detectors". *Physica C: Superconductivity*, **468**, 627 – 630, 2008.

[105] A. Verevkin, J. Zhang, R. Sobolewski, A. Lipatov, O. Okunev, G. Chulkova, A. Korneev, K. Smirnov, G. N. Gol'tsman, and A. Semenov. "Detection efficiency of large-active-area NbN single-photon superconducting detectors in the ultraviolet to near-infrared range". *Applied Physics Letters*, **80**, 4687–4689, 2002.

[106] A. Korneev, Y. Vachtomin, O. Minaeva, A. Divochiy, K. Smirnov, O. Okunev, G. Gol'tsman, C. Zinoni, N. Chauvin, L. Balet, F. Marsili, D. Bitauld, B. Alloing, L. Li, A. Fiore, L. Lunghi, A. Gerardino, M. Halder, C. Jorel, and H. Zbinden. "Single-Photon Detection System for Quantum Optics Applications". *Selected Topics in Quantum Electronics, IEEE Journal of*, **13**, 944 –951, july-aug. 2007.

[107] S. Adam, L. Piraux, S. Michotte, D. Lucot, and D. Mailly. "Stabilization of non-self-spreading hotspots in current- and voltage-biased superconducting NbN microstrips". *Superconductor Science and Technology*, **22**, 105010, 2009.

[108] L. N. Bulaevskii, M. J. Graf, and V. G. Kogan. "Vortex-assisted photon counts and their magnetic field dependence in single-photon superconducting detectors". *Physical Review B*, **85**, 014505, January 2012.

[109] A. N. Zotova and D. Y. Vodolazov. "Photon detection by current-carrying superconducting film: A time-dependent Ginzburg-Landau approach". *Phys. Rev. B*, **85**, 024509, Jan 2012.

[110] L. D. Landau and E. M. Lifsic. *Course of theoretical physics*, volume 5: Statistical physics. Pergamon Pr., Oxford, 2. ed., rev. and enl. edition, 1970.

[111] M. M. Kreitman. "Low Temperature Thermal Conductivity of Several Greases". *Review of Scientific Instruments*, **40**, 1562–1565, 1969.

[112] D. R. Smith and F. R. Fickett. "Low-Temperature Properties of Silver". *Journal of Research of the National Institute of Standards and Technology*, **100**, 119, 1995.

[113] V. V. Yurchenko, K. Ilin, J. M. Meckbach, M. Siegel, A. J. Qviller, Y. M. Galperin, and T. H. Johansen. "Thermo-magnetic stability of superconducting films controlled by nano-morphology". *Applied Physics Letters*, **102**, –, 2013.

[114] T. Yamashita, S. Miki, K. Makise, W. Qiu, H. Terai, M. Fujiwara, M. Sasaki, and Z. Wang. "Origin of intrinsic dark count in superconducting nanowire single-photon detectors". *Applied Physics Letters*, **99**, 161105, 2011.

[115] K. Likharev and V. Semenov. "RSFQ logic/memory family: a new Josephson-junction technology for sub-terahertz-clock-frequency digital systems". *Applied Superconductivity, IEEE Transactions on*, **1**, 3–28, 1991.

[116] P. Bunyk, K. Likharev, and D. Zinoviev. "RSFQ TECHNOLOGY: PHYSICS AND DEVICES". *International Journal of High Speed Electronics and Systems*, **11,1**, 257–305, 2001.

[117] O. Wetzstein. *Methoden zur Verbesserung der Funktionsstabilität digitaler Elektronik mit niedriger Schaltenergie*. PhD thesis, TU Ilmenau, 2010.

[118] T. Ortlepp and F. H. Uhlmann. "Technology Related Timing Jitter in Superconducting Electronics". *Applied Superconductivity, IEEE Transactions on*, **17**, 534 –537, june 2007.

[119] T. Ortlepp, O. Wetzstein, S. Engert, J. Kunert, and H. Toepfer. "Reduced Power Consumption in Superconducting Electronics". *Applied Superconductivity, IEEE Transactions on*, **21**, 770 –775, june 2011.

[120] T. Ortlepp, S. Wuensch, M. Schubert, P. Febvre, B. Ebert, J. Kunert, E. Crocoll, H.-G. Meyer, M. Siegel, and F. H. Uhlmann. "Superconductor-to-Semiconductor Interface Circuit for High Data Rates". *Applied Superconductivity, IEEE Transactions on*, **19**, 28 –34, feb. 2009.

[121] S. Anders, M. Blamire, F.-I. Buchholz, D.-G. Crete, R. Cristiano, P. Febvre, L. Fritzsch, A. Herr, E. Ilichev, J. Kohlmann, J. Kunert, H.-G. Meyer, J. Niemeyer, T. Ortlepp, H. Rogalla, T. Schurig, M. Siegel, R. Stolz, E. Tarte, H. ter Brake, H. Toepfer, J.-C. Villegier, A. Zagoskin, and A. Zorin. "European roadmap on superconductive electronics: status and perspectives". *Physica C: Superconductivity*, **470**, 2079 – 2126, 2010.

[122] A. Katara, A. Bapat, R. Chalse, A. Selokar, and S. Ramteke. "Development of One Bit Delta-Sigma Analog to Digital Converter". In *Computational Intelligence and Communication Networks (CICN), 2012 Fourth International Conference on*, pp. 300–304, 2012.

[123] T. Filippov and V. Kornev. "Sensitivity of the balanced Josephson-junction comparator". *Magnetics, IEEE Transactions on*, **27**, 2452–2455, 1991.

[124] T. Ortlepp, S. Whiteley, L. Zheng, X. Meng, and T. Van Duzer. "High-Speed Hybrid Superconductor-to-Semiconductor Interface Circuit With Ultra-Low Power Consumption". *Applied Superconductivity, IEEE Transactions on*, **23**, 1400104–1400104, 2013.

[125] O. Brandel, O. Wetzstein, T. May, H. Toepfer, T. Ortlepp, and H.-G. Meyer. "RSFQ electronics for controlling superconducting polarity switches". *Superconductor Science and Technology*, **25**, 125012, 2012.

[126] K. D. Irwin, M. D. Niemack, J. Beyer, H. M. Cho, W. B. Doriese, G. C. Hilton, C. D. Reintsema, D. R. Schmidt, J. N. Ullom, and L. R. Vale. "Code-division multiplexing of superconducting transition-edge sensor arrays". *Superconductor Science and Technology*, **23**, 034004, 2010.

[127] V. Ipatov. *Spread Spectrum and CDMA*. John Wiley & Sons, 2005.

[128] E.-L. Kuan, S. X. Ng, and L. Hanzo. "Joint-detection and interference cancellation based burst-by-burst adaptive CDMA schemes". *Vehicular Technology, IEEE Transactions on*, **51**, 1479 – 1493, nov 2002.

[129] Y. Eldar and A. Chan. "An optimal whitening approach to linear multiuser detection". *Information Theory, IEEE Transactions on*, **49**, 2156 – 2171, sept. 2003.

[130] S. Wuensch, G. Benz, E. Crocoll, M. Fitsilis, M. Neuhaus, T. Scherer, and W. Jutzi. "Normal and superconductor coplanar waveguides with 100 nm line width". *Applied Superconductivity, IEEE Transactions on*, **11**, 115 –118, mar 2001.

[131] J. C. Swihart. "Field Solution for a Thin-Film Superconducting Strip Transmission Line". *Journal of Applied Physics*, **32**, 461–469, 1961.

[132] M. Ejrnaes, R. Cristiano, O. Quaranta, S. Pagano, A. Gaggero, F. Mattioli, R. Leoni, B. Voronov, and G. Gol'tsman. "A cascade switching superconducting single photon detector". *Applied Physics Letters*, **91**, 262509, 2007.

[133] P. Marco. "Entwurf und Design eines integrierten BiasTee fuer HF-Auskopplung bei 4.2K". Master's thesis, KIT, 2011.

[134] K. Il'in, A. Stockhausen, A. Scheuring, M. Siegel, A. Semenov, H. Richter, and H.-W. Huebers. "Technology and Performance of THz Hot-Electron Bolometer Mixers". *Applied Superconductivity, IEEE Transactions on*, **19**, 269 –273, june 2009.

E Own publications

[CBC+14] M. Caselle, M. Balzer, S. Chilingaryan, M. Hofherr, V. Judin, A. Kopmann, N. J. Smale, P. Thoma, S. Wuensch, A. S. Müller, M. Siegel, and M. Weber. An ultra-fast data acquisition system for coherent synchrotron radiation with terahertz detectors. *Journal of Instrumentation*, 9(01):C01024, 2014.

[EIS+13] A. Engel, K. Inderbitzin, A. Schilling, R. Lusche, A. Semenov, H.-W. Hübers, D. Henrich, M. Hofherr, K. Il'in, and M. Siegel. Temperature-Dependence of Detection Efficiency in NbN and TaN SNSPD. *Applied Superconductivity, IEEE Transactions on*, 23(3):2300505–2300505, 2013.

[EWH+13] S. Engert, O. Wetzstein, M. Hofherr, K. Ilin, M. Siegel, H.-G. Meyer, and H. Toepfer. Mathematical Analysis of Multiplexing Techniques for SNSPD Arrays. *Applied Superconductivity, IEEE Transactions on*, PP(99):1–1, 2013.

[HAI+13] M. Hofherr, M. Arndt, K. Il'in, D. Henrich, M. Siegel, J. Toussaint, T. May, and H.-G. Meyer. Time-Tagged Multiplexing of Serially Biased Superconducting Single-Photon Detectors. *Applied Superconductivity, IEEE Transactions on*, PP(99):1–1, 2013.

[HDH+12] D. Henrich, S. Dörner, M. Hofherr, K. Il'in, A. Semenov, E. Heintze, M. Scheff-ler, M. Dressel, and M. Siegel. Broadening of hot-spot response spectrum of superconducting NbN nanowire single-photon detector with reduced nitrogen content. *Journal of Applied Physics*, 112(7):074511, 2012.

[HRD+13] D. Henrich, L. Rehm, S. Dorner, M. Hofherr, K. Il'in, A. Semenov, and M. Siegel. Detection Efficiency of a Spiral-Nanowire Superconducting Single-Photon Detector. *Applied Superconductivity, IEEE Transactions on*, 23(3):2200405–2200405, 2013.

[HRH+12] D. Henrich, P. Reichensperger, M. Hofherr, J. M. Meckbach, K. Il'in, M. Siegel, A. Semenov, A. Zotova, and D. Y. Vodolazov. Geometry-induced reduction of the critical current in superconducting nanowires. *Phys. Rev. B*, 86:144504, Oct 2012.

[HRI+10a] M. Hofherr, D. Rall, K. Ilin, A. Semenov, N. Gippius, H.-W. Hübers, and M. Siegel. Superconducting nanowire single-photon detectors: Quantum efficiency vs. film thickness. *Journal of Physics: Conference Series*, 234(1):012017, 2010.

[HRI⁺10b] M. Hofherr, D. Rall, K. Ilin, M. Siegel, A. Semenov, H.-W. Huebers, and N. A. Gippius. Intrinsic detection efficiency of superconducting nanowire single-photon detectors with different thicknesses. *Journal of Applied Physics*, 108(1):014507 –014507–9, jul 2010.

[HRI⁺12] M. Hofherr, D. Rall, K. Il'in, A. Semenov, H.-W. Hübers, and M. Siegel. Dark Count Suppression in Superconducting Nanowire Single Photon Detectors. *Journal of Low Temperature Physics*, pages 1–5, 2012. 10.1007/s10909-012-0495-9.

[HWE⁺12] M. Hofherr, O. Wetzstein, S. Engert, T. Ortlepp, B. Berg, K. Ilin, D. Henrich, R. Stolz, H. Toepfer, H.-G. Meyer, and M. Siegel. Orthogonal sequencing multiplexer for superconducting nanowire single-photon detectors with RSFQ electronics readout circuit. *Opt. Express*, 20(27):28683–28697, Dec 2012.

[IHR⁺11] K. Ilin, M. Hofherr, D. Rall, M. Siegel, A. Semenov, A. Engel, K. Inderbitzin, A. Aeschbacher, and A. Schilling. Ultra-thin TaN Films for Superconducting Nanowire Single-Photon Detectors. *Journal of Low Temperature Physics*, pages 1–6, 2011. 10.1007/s10909-011-0424-3.

[MTG⁺12] T. May, J. Toussaint, R. Grüner, M. Schubert, H.-G. Meyer, B. Dietzek, J. Popp, M. Hofherr, K. Il'in, D. Henrich, M. Arndt, and M. Siegel. Superconducting nanowire single-photon detectors for picosecond time resolved spectroscopic applications. *Biophotonics: Photonic Solutions for Better Health Care III*, 8427(1):84274B, 2012.

[OHF⁺11] T. Ortlepp, M. Hofherr, L. Fritzsch, S. Engert, K. Ilin, D. Rall, H. Toepfer, H.-G. Meyer, and M. Siegel. Demonstration of digital readout circuit for superconducting nanowire single photon detector. *Opt. Express*, 19(19):18593–18601, Sep 2011.

[OHT⁺13] T. Ortlepp, M. Hofherr, J. Toussaint, K. Ilin, H.-G. Meyer, and M. Siegel. Single flux quantum circuit for SNSPD readout with improved sensitivity and scalability for integrated multi-pixel arrays. *Invited talk on the 14th International Superconductive Electronics Conference (ISEC), Cambridge, MA, USA*, 2013.

[PSH⁺11] P. Probst, A. Scheuring, M. Hofherr, D. Rall, S. Wünsch, K. Il'in, M. Siegel, A. Semenov, A. Pohl, H.-W. Hübers, V. Judin, A.-S. Müller, A. Hoehl,

R. Müller, and G. Ulm. $YBa_2Cu_3O_{7-\delta}$ quasioptical detectors for fast time-domain analysis of terahertz synchrotron radiation. *Applied Physics Letters*, 98(4):043504, 2011.

[PSH+12] P. Probst, A. Scheuring, M. Hofherr, S. Wünsch, K. Ilin, A. Semenov, H.-W. Hübers, V. Judin, A.-S. Müller, J. Hänisch, B. Holzapfel, and M. Siegel. Superconducting $YBa_2Cu_3O_{7-\delta}$ Thin Film Detectors for Picosecond THz Pulses. *Journal of Low Temperature Physics*, 167(5-6):898–903, 2012.

[RHH+13] L. Rehm, D. Henrich, M. Hofherr, S. Wuensch, P. Thoma, A. Scheuring, K. Il'in, M. Siegel, S. Haindl, K. Iida, F. Kurth, B. Holzapfel, and L. Schultz. Infrared Photo-Response of Fe-Shunted Ba-122 Thin Film Microstructures. *Applied Superconductivity, IEEE Transactions on*, PP(99):1–1, 2013.

[RPH+10] D. Rall, P. Probst, M. Hofherr, S. Wünsch, K. Il'in, U. Lemmer, and M. Siegel. Energy relaxation time in NbN and YBCO thin films under optical irradiation. *Journal of Physics: Conference Series*, 234(4):042029, 2010.

[TGS+12] J. Toussaint, R. Grüner, M. Schubert, T. May, H.-G. Meyer, B. Dietzek, J. Popp, M. Hofherr, M. Arndt, D. Henrich, K. Il'in, and M. Siegel. Superconducting single-photon counting system for optical experiments requiring time-resolution in the picosecond range. *Review of Scientific Instruments*, 83(12):123103, 2012.

[TRS+13] P. Thoma, J. Raasch, A. Scheuring, M. Hofherr, K. Il'in, S. Wunsch, A. Semenov, H.-W. Hubers, V. Judin, A.-S. Muller, N. Smale, J. Hanisch, B. Holzapfel, and M. Siegel. Highly Responsive Y -Ba-Cu -O Thin Film THz Detectors With Picosecond Time Resolution. *Applied Superconductivity, IEEE Transactions on*, 23(3):2400206–2400206, June 2013.

F Supervised student theses

1. Ferdinand Schwenk, "Optimiertes TCSPC Verfahren für die Auslese von Time-tagged SNSPD Arrays."

2. Robert Korn, "Development of a FPGA based high-resolving time-to-digital converter for fast SNSPD readout."

3. Marco Neber, "Kalibrierte Temperaturmessungen für kryogene Systeme".

4. Dorjan Sulaj, "Automatisierung von mobilen Biasquellen für die Detektorcharakterisierung: Datenverwaltung und Signalbusansteuerung."'

5. Hanna Becker, "Automatisierung von mobilen Biasquellen für die Detektorcharakterisierung: digital/analog Wandlung und Softwareansteuerung."'

6. Constantin Becker, "Entwicklung eines rausch- und störarmen Leistungsnetzteils für die IMS-Messplattform."

7. Benjamin Berg, "Implementierung einer verbesserten Grauwertauflösung des FDM-Auslesesystems."

8. Matthias Arndt, "Entwicklung eines Mischsignal-Messmoduls für Jittermessungen im Picosekundenbereich."

9. Alexei Witzig, "Entwicklung einer einsatzfähigen Systemperipherie für eine Multisensor-Messplattform."

10. Bao An, "Entwurf eines FPGA basierten SPI-Busses zur Ansteuerung der IMS Messplattform."

11. Pavol Marco, "Entwurf und Design eines integrierten BiasTee fuer HF-Auskopplung bei 4.2 K."

12. Ferdinand Schwenk, "Optimierung eines Frequenzausleseverfahrens für supraleitende Multipixeldetektoren."

13. Robert Prinz, "Entwicklung und Untersuchung von LEKIDs für den Aufbau eines 1k-pixel Arrays."

14. Valentin Ndasse, "Entwicklung eines elektronischen Auslesesystems für Single Photon Detektoren (SSPD)."

15. Leo Gosslich, "Entwicklung eines adaptiven Frequenzausleseverfahrens für supraleitende Multipixeldetektoren."

16. Julia Toussaint, "Kryomesssystem für die CARS-Spektroskopie."

17. Jörg Schaaf, "Entwurf und Design einer präzisen Temperaturregelung für kryogene Messungen."

18. Matthias Arndt, "Entwurf und Aufbau einer rauscharmen Bias-Stromquelle für supraleitende Detektoren."

19. Markus Ditlevsen, "Entwurf und Aufbau einer Höchstfrequenz-Ansteuerelektronik für eine VCSEL-Laserdiode."

G Contributions to international conferences

1. M. Hofherr, J. Toussaint, M. Arndt, K. Ilin, T. May, H.-G. Meyer, M. Siegel; Time-tagged multiplexing of SNSPD arrays for pulsed and cw applications, Kryoelektronische Bauelemente 2013

2. T. Ortlepp, M. Hofherr, K. Ilin, M. Siegel, Single flux quantum circuit for SNSPD readout with improved sensitivity and scalability for integrated multipixel arrays, Kryoelektronische Bauelemente 2013

3. J. Toussaint, T. May, H.-G. Meyer, M. Hofherr, K. Ilin, M. Siegel, Potentials and limitations of Superconsucting Nanowire Single-Photon Detectors for time-resolved measurements, Kryoelektronische Bauelemente 2013

4. (INVITED) T. Ortlepp, M. Hofherr, O. Brandel, J. Toussaint, K. Ilin, T. May, The path towards multi-pixel arrays of superconducting detectors, Biennial European Conference on Applied Superconductivity, Genua, Italy (2013)

5. M. Hofherr, J. Toussaint, M. Arndt, K. Ilin, T. May, H.-G. Meyer, M. Siegel, Time-tagged multiplexing of SNSPDs for multi-pixel arrays, Biennial European Conference on Applied Superconductivity, Genua, Italy (2013)

6. (INVITED) T. Ortlepp, M. Hofherr, K. Ilin, J. Toussaint, M. Siegel, Single flux quantum circuit for SNSPD readout with improved sensitivity and scalability for integrated multi-pixel arrays, 14th International Superconductivity Electronics Conference (2013)

7. P. Trojan, K. Ilin, D. Henrich, M. Hofherr, A. Semenov, H.-W. Hübers and M. Siegel, Superconducting nanowire single-photon detectors (SNSPDs) on SOI for near-infrared range, DPG-Frühjahrsragung, Regensburg, Germany (2013)

8. P. Probst, A. Scheuring, M. Hofherr, S. Wuensch, K. Ilin, A. Pohl, A. Semenov, H.-W. Huebers, V. Judin, A.-S. Mueller, A. Hoehl, R. Mueller, G. Ulm, J. Haenisch, B. Holzapfel and M. Siegel, $YBa_2Cu_3O_{7-\delta}$ high-speed detectors for picosecond THz pulses, 23rd International Symposium on Space Terahertz Technology, Tokyo, 2-4 April 2012

9. M. Hofherr, K. Il'in, D. Henrich, M. Arndt, M. Siegel, J. Toussaint, R. Grüner, T. May, M. Schubert, H.-G. Meyer, B. Dietzek, J. Popp, Superconducting nanowire single-photon detectors for time correlated single photon counting, FLIM, Saarbrücken (2012)

10. M. Hofherr, O. Wetzstein, S. Engert, T. Ortlepp, D. Rall, K. Ilin, R. Stolz, H. Töpfer, H.-G. Meyer, M: Siegel, Code division multiplexing for superconducting nanowire single photon detectors using RSFQ electronic, Applied Superconductivity Conference, Portland, USA (2012)

11. M. Hofherr, K. Ilin, D. Henrich, S. Wünsch, M. Siegel, J. Toussaint, M. Schubert, T. May, H.-G. Meyer, R. Grüner, B. Dietzek, Superconducting nanowire single-photon detectors for picosecond time resolved spectroscopic applications, Applied Superconductivity Conference, Portland, USA (2012)

12. S. Engert, O. Wetzstein, M. Hofherr, K. Ilin, M. Siegel, H.-G. Meyer, H. Toepfer, Concept study of a multiplexing technique for superconducting nanowire single-photon detector arrays, Applied Superconductivity Conference, Portland, USA (2012)

13. S. Engert, O. Wetzstein, M. Hofherr, K. Ilin, H.-G. Meyer, H. Toepfer, Concept study to derive parameter limits for orthogonal sequencing multiplexing for SNSPD arrays, Meeting on kryoelectronic devices 2011, Autrans (Grenoble), France (2011)

14. J. Toussaint, M. Hofherr, D. Henrich, K. Il'in, A. Semenov, H.-W. Hübers, H.-G. Meyer, M. Siegel, Superconducting nanowire single-photon detectors: challenges in optimization of detector performance Meeting on kryoelectronic devices 2011, Autrans (Grenoble), France (2011)

15. O. Wetzstein, M. Hofherr, S. Engert, L. Fritzsch, T. Ortlepp, K. Il'in, D. Rall, R. Stolz, H. Töpfer, M. Siegel, H.-G. Meyer, Simultaneous readout of a multi-pixel superconducting single-photon detector array, Meeting on kryoelectronic devices 2011, Autrans (Grenoble), France (2011)

16. M. Hofherr, D. Henrich, K. Il'in, J. Toussaint, A. Semenov, H.-W. Hübers, H.-G. Meyer, M. Siegel, Superconducting nanowire single-photon detectors: challenges in optimization of detector performance, 7th FLUXONICS RSFQ design workshop, Ilmenau, Germany (2011)

17. O. Wetzstein, M. Hofherr, S. Engert, L. Fritzsch, T. Ortlepp, K. Il'in, D. Rall, R. Stolz, H. Töpfer, M. Siegel, H.-G. Meyer, Simultaneous readout of a multi-pixel superconducting single-photon detector array, 7th FLUXONICS RSFQ design workshop, Ilmenau, Germany (2011)

18. J. Toussaint, M. Hofherr, M. Schubert, K. Il'in, D. Rall, S. Wünsch, M. Siegel, H.-G. Meyer, T. May, R. Grüner, D. Akimov, B. Dietzek, Superconducting micro-bolometers and single photon detectors for spectroscopic applications, Meeting on kryoelectronic devices 2011, Autrans (Grenoble), France (2011)

19. A. Engel, K. Inderbitzin, A. Aeschbacher, A. Schilling, D. Rall, M. Hofherr, K. Il'in, M. Siegel, Detection of X-rays using Superconducting Nanowire Single-Photon Detectors, International Workshop on Nanowire Superconducting Single-Photon Detectors, Eindhoven, the Netherlands (2011)

20. M. Hofherr, K. Il'in, D. Rall, A. Semenov, H. Bartolf, A. Engel, A. SChilling, H.-W. Hübers, M. Siegel, Suppression of dark counts of superconducting nanowire single-photon detectors caused by thermal fluctuations, Superconductivity Centennial Conference EUCAS-ISEC-ICMC, World Forum, The Hague, the Netherlands (2011)

21. M. Hofherr, K. Il'in, D. Rall, O. Wetzstein, L. Fritsch, S. Engert, T. Ortlepp, A. Semenov, H.-W. Hübers, H.-G. Meayer, H. Töpfer, M. Siegel, Single-flux quantum multiplexer for multi-pixel superconducting single-photon detector arrays, Superconductivity Centennial Conference EUCAS-ISEC-ICMC, World Forum, The Hague, the Netherlands (2011)

22. K. Il'in, D. Rall, M. Hofherr, M. Siegel, A. Semenov, H.-W. Hübers, A. Engel, K. Inderbitzin, A. Aeschbacher, A. Schilling, Ultra-thin TaN film nanowires for single-photon detection, Superconductivity Centennial Conference EUCAS-ISEC-ICMC, World Forum, The Hague, the Netherlands (2011)

23. D. Rall, J. Toussaint, M. Hofherr, S. Wuensch, K. S. Ilin, A. Semenov, H.-W. Huebers, M. Siegel, Detection Efficiency and Cut-off Wavelength of NbN-SNSPD with Different Stoichiometry, 14th International Workshop on low temperature detectors, Heidelberg, Germany (2011)

24. M. Hofherr, K. Ilin, A. Semenov, D. Rall, H. Bartolf, A. Engel, A. Schilling, H.-W. Hübers, M. Siegel, Fluctuations in thin NbN Nanowire Single-Photon Detectors, 14th International Workshop on low temperature detectors, Heidelberg, Germany (2011)

25. P. Jung, A. Lukashenko, A. Zhuravel, S.Wuensch, M. Hofherr, K. Ilin, M. Siegel, and A. Ustinov, Spatially-Resolved Single-Photon Detection in NbN Nanowires, DPG Frühjahrstagung, Dresden, Germany (2011)

26. K. Ilin, D. Rall, M. Hofherr, M. Siegel, A. Semenov, H.-W. Hübers, A. Engel, K. Inderbitzin, A. Aeschbacher, A. Schilling, Superconducting ultra-thin TaN film nanostructures for optical and infrared detectors, 14th International Workshop on low temperature detectors, Heidelberg, Germany (2011)

27. (INVITED) T. Ortlepp, M. Hofherr, K. Il'in, S. Engert, D. Rall, S. Wuensch, H. Toepfer, M. Siegel, RSFQ based readout of superconducting single-photon detectors, Applied Superconductivity Conference 2010, Washington, USA (2010)

28. D. Rall, J. Toussaint, M. Hofherr, S. Wuensch, K. Ilin, A. Semenov, H.-W. Huebers, U. Lemmer and M. Siegel, Improvement of SNSPD Detection Efficiency by Variation of NbN Chemical Composition, Tagung kryoelektronische Bauelemente, Zeuthen, Germany (2010)

29. T. Ortlepp, M. Hofherr, K. Il'in, S. Engert, D. Rall, H. Toepfer, M. Siegel, RSFQ based readout of superconducting single photon detectors Tagung kryoelektronische Bauelemente, Zeuthen, Germany (2010)

30. K. Il'in, D. Rall, M.Hofherr, M. Siegel, A. Semenov, H.-W. Hübers, A. Engel A. Schilling, TaN thin films for Superconducting Nanowire Single-Photon Detectors Tagung kryoelektronische Bauelemente, Zeuthen, Germany (2010)

31. P. Probst, A. Scheuring, M. Hofherr, A. Stockhausen, D. Rall, S. Wuensch, K. Ilin, A. Semenov, H.-W. Hübers, V. Judin, A.-S. Müller, J. Hänisch, B. Holzapfel and M. Siegel, High-speed YBCO detectors for resolving picosecond THz pulses, Tagung kryoelektronische Bauelemente, Zeuthen, Germany (2010)

32. D. Rall, K. S. Ilin, M. Hofherr, P. Probst, S. Wuensch, U. Lemmer, M. Siegel, A. D. Semenov, H.-W. Huebers, Investigation of NbN thin film response to optical irradiation for detector application, 2nd S-Pulse Karlsruhe Detector Workshop, Karlsruhe, Germany (2010)

33. M. Hofherr, K. Ilin, A. Semenov, N. Gippius, D. Rall, H.- W. Huebers, M. Siegel, Improved understanding of superconducting nanowire single-photon detectors, a key to a better performance, KHYSSymposium (2010)

34. M. Hofherr, K. Ilin, D. Rall, M. Siegel, A. D. Semenov, H.-W. Huebers, N. Gippius "Intrinsic detection efficiency of superconducting nanowire single-photon detectors with different thicknesses" 2nd S-Pulse Karlsruhe Detector Workshop, Karlsruhe, Germany (2010)

35. P. Probst, D. Rall, A. Scheuring, M. Hofherr, S. Wuensch, K. Ilin, A. Semenov, H.-W. Huebers, V. Judin, A.-S. Mueller and M. Siegel, Superconducting YBCO hot-electron bolometer detectors for picosecond time resolution of THz pulses 456. Wilhelm und Else Heraeus Seminar, THz radiation: Generation, Detection and Applications, Bad Honnef, Germany (2010)

36. P. Probst, D. Rall, M. Hofherr, S. Wünsch, K. Ilin and M. Siegel, Energy relaxation processes in YBCO thin films studied by frequency and time-domain techniques, DPG Frühjahrstagung der Sektion Kondensierte Materie (SKM), Regensburg, Germany (2010)

37. P. Jung, A. Lukashenko, A. Zhuravel, S. Wuensch, M. Hofherr, K. Ilin, M. Siegel, and A. Ustinov, Spacially resolved investigation of superconducting NbN nanowire single photon detectors, DPG Frühjahrstagung der Sektion Kondensierte Materie (SKM), Regensburg, Germany (2010)

38. M. Hofherr, K. Ilin, A. Semenov, N. Gippius, D. Rall, H.- W. Huebers, M. Siegel Superconducting nanowire single-photon detectors: Intrinsic quantum efficiency vs. thickness of NbN films, Tagung Kryoelektronische Bauelemente, Oberhof, Germany (2009)

39. P. Probst, D. Rall, M. Hofherr, S. Wünsch, K. Ilin and M. Siegel, Measurements of energy relaxation time in thin $YBa_2Cu_3O_{7-\delta}$ films on sapphire substrates, Kryoelektronische Bauelemente 2009, Oberhof, Germany (2009)

40. M. Hofherr, K. Ilin, A. Semenov, N. Gippius, D. Rall, H.- W. Huebers, M. Siegel Superconducting nanowire single-photon detectors: Intrinsic quantum efficiency vs. thickness of NbN films, European Conference on Applied Superconductivity, Dresden, Germany (2009)

41. D. Rall, P. Probst, M. Hofherr, S. Wünsch, K. Ilin, U. Lemmer, M. Siegel, Energy Relaxation Time in NbN and $YBa_2Cu_3O_{7-\delta}$ Thin Films under Optical Radiation, European Conference on Applied Superconductivity, Dresden, Germany (2009)

42. D. Rall, M. Hofherr, K. Il'in, S. Wünsch, M. Siegel, U. Lemmer, A. Semenov, H.- W. Huebers, Investigation of Energy Relaxation Processes in NbN thin films using Optical Irradiation, DPG Frühjahrstagung der Sektion Kondensierte Materie (SKM), Dresden, Germany, (2009)

Danksagung

Cum Donum Necessarium Iuvat

Frei übersetzt, "Weil ein notwendig Geschenk hilfreich ist", darum ist eine Danksagung für die Unterstützung, die mir während meiner Dissertation zu Teil wurde, nur ein Ansatz etwas an Dank dafür zurückzugeben. Ich spreche von einem Geschenk, da Support nicht als selbstverständlich wahrgenommen werden soll und von notwendig, da es zum Gelingen dieser Arbeit dennoch unverzichtbar war.

Ich möchte daher meinen Dank zuerst an meinen Doktorvater Herrn Prof. rer. nat. habil. Michael Siegel richten, der mir die Dissertation überhaupt erst ermöglicht hat, indem er meine Betreuung übernommen hat, mir Raum, Mittel und Freiheit für meine Forschung gegeben hat und auch viele Möglichkeiten, Ergebnisse international zu präsentieren.
Ich bedanke mich auch bei Prof. Dr.-Ing. habil. Hannes Töpfer von der TU Ilmenau für die Übernahme des Zweitgutachtens.
Herrn Dr. Konstantin Ilin möchte ich herzlich danken für seine direkte Betreuung, seinen immer hilfreichen und geduldigen Rat und seine Unterstützung beim Detektorbau.
Diese Arbeit wurde in enger Zusammenarbeit mit diversen Instituten durchgeführt. Daher gilt auch mein Dank an all diejenigen, die mir das ermöglicht haben. Dazu gehören Prof. Alexei Semenov und Prof. Heinz-Wilhelm Hübers vom DLR Berlin, Institut für Planetenkunde, Prof. Hans-Georg Meyer, Torsten May und Marco Schubert vom IPHT Jena, Abteilung für Quantendetektion, Dr. Thomas Ortlepp vom CIS Erfurt und Prof. Hannes Töpfer, TU Ilmenau, Department of Advanced Electromagnetics.
Es ist schwierig, den einzelnen weiteren Leuten im Dank gerecht zu werden, ohne dabei irgendetwas zu vergessen. Jeder hat auf seine Art individuell und unverzichtbar zum Erfolg dieser Arbeit beigetragen. Ich möchte daher nicht im Detail auf die einzelnen Tätigkeiten eingehen. Aber ich möchte die Namen nennen:
Bao An, Matthias Arndt, Constantin Becker, Hanna Becker, Benjamin Berg, Björn Brenner, Dr. Michele Caselle, Erich Crocoll, Markus Ditlevsen, Doris Duffner, Sonja Engert, Lea Fuchs, N.A. Gippius, Leo Gosslich, Karlheinz Gutbrodt, Dr. Gerd Hammer, Dr. Dagmar Henrich, Dr. Christoph Kaiser, Matthias Kohler, Robert Korn, Letizia Macri, Pavol Marko, Dr. Max Meckbach, Michael Merker, Valentin Ndasse, Marco Neber, Robert Prinz, Juliane Raasch, Jörg Schaaf, Dr. Alexander Scheuring, Ferdinand Schwenk, Alexander Stassen, Dr. Axel Stockhausen, Dorjan Sulaj, Dr. Petra Thoma, Julia Toussaint, Phillip Trojan, Hansjürgen Wermund, Olaf Wetzstein, Alexei Witzig, Dr. Stefan Wünsch.
Des Weiteren gibt es immer passive Helfer, die durch ihr Erkundigen und Nachfragen durch

ihr "daran denken" oder mich "ablenken" ebenfalls motivationsfördernd für mich und diese Arbeit waren: Daher gilt der Dank allen meinen Freunden.

Zuletzt gebührt der Dank meiner lieben Familie, meiner Mutter und meinem Vater, meinen Brüdern und allen, die dazu gehören.

Matthias Hofherr
Karlsruhe, Dezember 2013

Karlsruher Schriftenreihe zur Supraleitung
(ISSN 1869-1765)

Herausgeber: Prof. Dr.-Ing. M. Noe, Prof. Dr. rer. nat. M. Siegel

Die Bände sind unter www.ksp.kit.edu als PDF frei verfügbar oder als Druckausgabe bestellbar.

Band 001
Christian Schacherer
Theoretische und experimentelle Untersuchungen zur Entwicklung supraleitender resistiver Strombegrenzer. 2009
ISBN 978-3-86644-412-6

Band 002
Alexander Winkler
Transient behaviour of ITER poloidal field coils. 2011
ISBN 978-3-86644-595-6

Band 003
André Berger
Entwicklung supraleitender, strombegrenzender Transformatoren. 2011
ISBN 978-3-86644-637-3

Band 004
Christoph Kaiser
High quality Nb/Al-AlOx/Nb Josephson junctions. Technological development and macroscopic quantum experiments. 2011
ISBN 978-3-86644-651-9

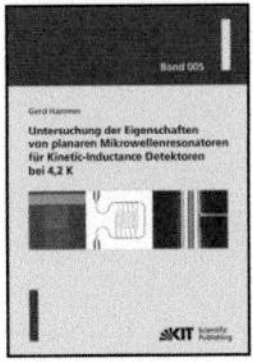

Band 005
Gerd Hammer
Untersuchung der Eigenschaften von planaren Mikrowellen-resonatoren für Kinetic-Inductance Detektoren bei 4,2 K. 2011
ISBN 978-3-86644-715-8

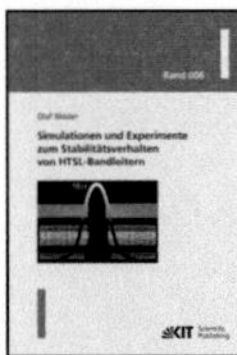

Band 006
Olaf Mäder
**Simulationen und Experimente zum Stabilitätsverhalten
von HTSL-Bandleitern.** 2012
ISBN 978-3-86644-868-1

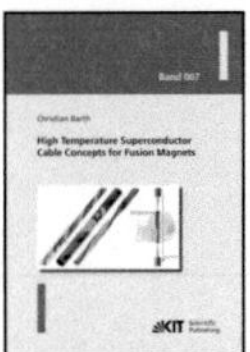

Band 007
Christian Barth
**High Temperature Superconductor Cable Concepts for
Fusion Magnets.** 2013
ISBN 978-3-7315-0065-0

Band 008
Axel Stockhausen
Optimization of Hot-Electron Bolometers for THz Radiation. 2013
ISBN 978-3-7315-0066-7

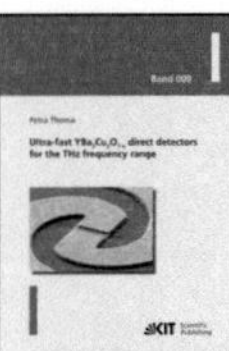

Band 009
Petra Thoma
**Ultra-fast YBa$_2$Cu$_3$O$_{7-x}$ direct detectors for the THz frequency
range.** 2013
ISBN 978-3-7315-0070-4

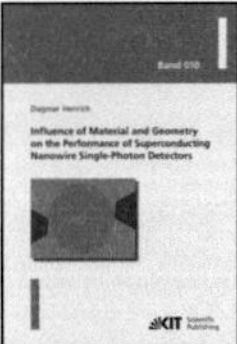

Band 010
Dagmar Henrich
**Influence of Material and Geometry on the Performance
of Superconducting Nanowire Single-Photon Detectors.** 2013
ISBN 978-3-7315-0092-6

Band 011
Alexander Scheuring
Ultrabreitbandige Strahlungseinkopplung in THz-Detektoren. 2013
ISBN 978-3-7315-0102-2

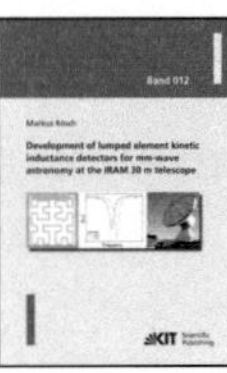

Band 012
Markus Rösch
**Development of lumped element kinetic inductance detectors
for mm-wave astronomy at the IRAM 30 m telescope.** 2013
ISBN 978-3-7315-0110-7

Band 013
Johannes Maximilian Meckbach
**Superconducting Multilayer Technology for Josephson
Devices.** 2013
ISBN 978-3-7315-0122-0

Band 014
Enrico Rizzo
**Simulations for the optimization of High Temperature Super-
conductor current leads for nuclear fusion applications.** 2014
ISBN 978-3-7315-0132-9

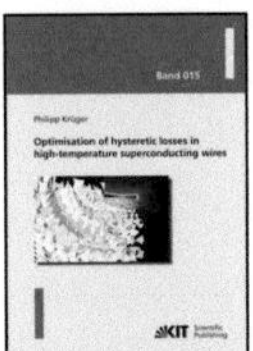

Band 015
Philipp Krüger
**Optimisation of hysteretic losses in high-temperature
superconducting wires.** 2014
ISBN 978-3-7315-0185-5

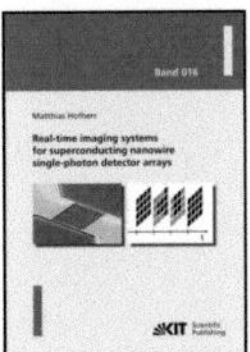

Band 016
Matthias Hofherr
**Real-time imaging systems for superconducting nanowire
single-photon detector arrays.** 2014
ISBN 978-3-7315-0229-6